D1168152

The Monograph

The volumes within the *Tall Buildings and Urban Environment* series correspond to the Council's group and committee structure. The present listing includes all current topical committees. Some are collaborating to produce volumes together, and Groups DM and BSS plan, with only a few exceptions, to combine all topics into one volume.

PLANNING AND ENVIRONMENTAL CRITERIA (PC)
Philosophy of Tall Buildings
History of Tall Buildings
Architecture
Rehabilitation, Renovation, Repair
Urban Planning and Design
External Transportation
Parking
Social Effects of the Environment
Socio-Political Influences
Design for the Disabled and Elderly
Interior Design
Landscape Architecture

DEVELOPMENT AND MANAGEMENT (DM)
Economics
Ownership and Maintenance
Project Management
Tall Buildings in Developing Countries
Decision-Making Parameters
Development and Investment
Legal Aspects

SYSTEMS AND CONCEPTS (SC)
Cladding
Partitions, Walls, and Ceilings
Structural Systems
Foundation Design
Construction Systems
High-Rise Housing
Prefabricated Tall Buildings
Tall Buildings Using Local Technology
Robots and Tall Buildings
Application of Systems Methodology

CRITERIA AND LOADING (CL)
Gravity Loads and Temperature Effects
Earthquake Loading and Response
Wind Loading and Wind Effects
Fire
Accidental Loading
Safety and Quality Assurance
Motion Perception and Tolerance

TALL STEEL BUILDINGS (SB)
Commentary on Structural Standards
Methods of Analysis and Design
Stability
Design Methods Based on Stiffness
Fatigue Assessment & Ductility Assurance
Connections
Cold-Formed Steel
Load and Resistance Factor Design (Limits States Design)
Mixed Construction

TALL CONCRETE AND MASONRY BUILDINGS (CB)
Commentary on Structural Standards
Selection of Structural Systems
Optimization
Elastic Analysis
Nonlinear Analysis and Limit Design
Stability
Stiffness and Crack Control
Precast Panel Structures
Creep, Shrinkage, & Temperature Effects
Cast-in-Place Concrete
Precast-Prestressed Concrete
Masonry Structures

BUILDING SERVICE SYSTEMS (BSS)
HVAC/Energy Conservation
Plumbing and Fire Protection
Electrical Systems
High-Tech Buildings
Vertical & Horizontal Transportation
Environmental Design
Urban Services

The basic objective of the Council's Monograph is to document the most recent developments to the state of the art in the field of tall buildings and their role in the urban habitat. The following volumes can be ordered through the Council.

Planning and Design of Tall Buildings, 5 volumes (1978-1981 by ASCE)

Developments in Tall Buildings–1983 (Van Nostrand Reinhold Company)

Advances in Tall Buildings (1986, Van Nostrand Reinhold Company)

High-Rise Buildings: Recent Progress (1986, Council on Tall Buildings)

Second Century of the Skyscraper (1988 Van Nostrand Reinhold Company)

Tall Buildings: 2000 and Beyond, 2 volumes (1990 & 1991, Council on Tall Buildings)

Council Headquarters
Lehigh University, Building 13
Bethlehem, Pennsylvania 18015 USA

Cold-Formed Steel in Tall Buildings

Library of Congress Cataloging-in-Publication Data

Cold-formed steel in tall buildings / Council on Tall Buildings and Urban Habitat, Committee S37 ; contributors, Rolf Baehre . . . [et al.] ; editorial group, Wei-Wen Yu, chairman, Rolf Baehre, vice -chairman, Ton Tomà, editor.
p. cm. — (Tall steel buildings) (Tall buildings and urban environment series)
Includes bibliographical references and index.
ISBN 0-07-012529-5
1. Building, Iron and steel. 2. Tall buildings—Design and construction. 3. Steel, Structural. I. Baehre, R. (Rolf) (date). II. Yu, Wei-wen (date). III. Tomà, Ton. IV. Council on Tall Buildings and Urban Habitat. Committee S37 (Cold-Formed Steel) V. Series. VI. Series: Tall buildings and urban environment series. VII. Series: Monograph (Council on Tall Buildings and Urban Habitat)
TA684.C63 1993
624.1′821—dc20 92-44849
CIP

ISBN 0-07-012529-5

1 2 3 4 5 6 7 8 9 0 DOC/DOC 9 9 8 7 6 5 4 3

For the Council on Tall Buildings, Lynn S. Beedle is the Editor-in-Chief and Dolores B. Rice is the Managing Editor.

For McGraw-Hill, the sponsoring editor was Joel Stein, the editing supervisor was Peggy Lamb, and the production supervisor was Pamela A. Pelton. This book was set in Times Roman by The Universities Press (Belfast) Ltd.
Printed and bound by R. R. Donnelley & Sons Company.

This book is printed on acid-free paper.

Council on Tall Buildings and Urban Habitat

Steering Group

Executive Committee

Members

Editorial Committee

Council on Tall Buildings and Urban Habitat

Sponsoring Societies

American Society of Civil Engineers (ASCE)
International Association for Bridge and Structural Engineering (IABSE)
American Institute of Architects (AIA)
American Planning Association (APA)
International Union of Architects (UIA)
American Society of Interior Designers (ASID)
Japan Structural Consultants Association (JSCA)
Urban Land Institute (ULI)
International Federation of Interior Designers (IFI)

The following identifies those firms and organizations who provide for the Council's financial support.

Patrons

Al-Rayes Group, Kuwait
Jaros, Baum & Bolles, New York
Kuwait Foundation for the Advancement of Sciences, Kuwait
National Science Foundation, Washington, DC
The Turner Corporation, New York
U.S. Agency for International Development, Washington, DC

Sponsors

Office of Irwin G. Cantor, P.C., New York
CMA Architects & Engineers, San Juan
Connell Wagner Rankine & Hill, Neutral Bay
Europrofile Tecom, Luxembourg
George A. Fuller Co., New York
Hong Kong Land Property Co. Ltd., Hong Kong
Johnson Controls, Inc., Milwaukee
Kohn, Pedersen, Fox Assoc., P.C., New York
Kone Elevators, Helsinki
Land Development Corporation, Hong Kong
John A. Martin & Assoc., Inc., Los Angeles
Ahmad Moharram, Cairo
Walter P. Moore & Associates, Inc., Houston
Nippon Steel, Tokyo
Ove Arup Partnership, London
PDM Strocal Inc., Stockton
Leslie E. Robertson Associates, New York
Schindler Management A.G., Luzerne
Swire Properties Ltd., Hong Kong
Takenaka Corporation, Tokyo
Tishman Construction Corporation of New York, New York
Tishman Speyer Properties, New York
Weiskopf & Pickworth, New York
Wing Tai Construction & Engineering, Hong Kong

Donors

American Bridge Co., Pittsburgh
American Iron and Steel Institute, Washington, D.C.
T. R. Hamzah & Yeang Sdn Bhd, Selangor
Haseko Corporation, Tokyo
The Herrick Corp., Pleasanton
Hollandsche Beton Maatschappij BV, Rijswijk
Hong Kong Housing Authority, Hong Kong
Iffland Kavanagh Waterbury, P.C., New York
Jacobs Brothers Co., Cleveland
Notch + Associates, Arlington
RTKL Associates, Inc., Baltimore
Skidmore, Owings & Merrill, Chicago
Steen Consultants Pty Ltd., Singapore
Syska & Hennesy, Inc., New York
Thornton-Tomasetti/Engineers, New York
Werner Voss & Partners, Braunschweig
Wong Hobach Lau Consulting Engineers, Los Angeles
Wong & Ouyang (HK) Ltd., Hong Kong

Contributors

Boundary Layer Wind Tunnel Laboratory (U. Western Ontario), London
H. K. Cheng & Partners Ltd., Hong Kong
Douglas Specialist Contractors Ltd., Aldridge
The George Hyman Construction Co., Bethesda
Ingenieurburo Muller Marl GmbH, Marl
INTEMAC, Madrid
Johnson Fain and Pereira Assoc., Los Angeles
The Kling-Lindquist Partnership, Inc., Philadelphia
LeMessurier Consultants Inc., Cambridge
W. L. Meinhardt & Partners Pty. Ltd., Melbourne
Obayashi Corporation, Tokyo
OTEP Internacional, SA, Madrid
PSM International, Chicago
Tooley & Company, Los Angeles
Nabih Youssef and Associates, Los Angeles

Contributing Participants

Adviesbureau Voor Bouwtechniek BV, Arnhem
American Institute of Steel Construction, Chicago
Anglo American Property Services (Pty.) Ltd., Johannesburg
Architectural Servies Dept., Hong Kong
Artech, Inc., Taipei
Atelier D'Architecture de Genval, Genval
Austin Commercial, Inc., Dallas
Australian Institute of Steel Construction, Milsons Point
B.C.V. Progetti S.r.l., Milano
Bechtel Corporation, San Francisco
W. S. Bellows Construction Corp., Houston
Alfred Benesch & Co., Chicago
BMP Consulting Engineers, Hong Kong
Bornhorst & Ward Pty. Ltd., Spring Hill
Bovis Limited, London
Bramalea Ltd., Dallas
Brandow & Johnston Associates, Los Angeles
Brooke Hillier Parker, Hong Kong
Campeau Corp., Toronto
CBM Engineers, Houston
Cermak Peterka Petersen, Inc., Fort Collins
Construction Consulting Laboratory, Dallas
Crane Fulview Door Co., Lake Bluff
Crone & Associates Pty. Ltd., Sydney
Crow Construction Co., New York
Davis Langdon & Everest, London
DeSimone, Chaplin & Dobryn, New York
Dodd Pacific Engineering, Inc., Seattle
Englekirk, Hart, and Sobel, Inc., Los Angeles
Falcon Steel Company, Wilmington
Fujikawa Johnson and Associates, Chicago
Gutteridge Haskins & Davey Pty. Ltd., Sydney
Hayakawa Associates, Los Angeles
Healthy Buildings Intl., Inc., Fairfax
Hellmuth, Obata & Kassabaum, Inc., San Francisco
Honeywell, Inc., Minneapolis
International Iron & Steel Institute, Brussels
Irwin Johnston and Partners, Sydney
JATOCRET, S.A., Rio de Janeiro
Johnson & Nielsen, Irvine
KPFF Consulting Engineers, Seattle
Lend Lease Design Group Ltd., Sydney
Stanley D. Lindsey & Assoc., Nashville
Lohan Associates, Inc., Chicago
Martin & Bravo, Inc., Honolulu
Enrique Martinez-Romero, S.A., Mexico
McWilliam Consulting Engineers, Brisbane
Mitchell McFarlane Brentnall & Partners Intl. Ltd., Hong Kong
Mitsubishi Estate Co., Ltd., Tokyo
Moh and Associates, Inc., Taipei
Mueser Rutledge Consulting Engineers, New York
Multiplex Constructions (NSW) Pty. Ltd., Sydney
Nihon Sekkei, U.S.A., Ltd., Los Angeles
Nikken Sekkei Ltd., Tokyo
Norman Disney & Young, Brisbane
O'Brien-Kreitzberg & Associates, Inc., Pennsauken
Ove Arup & Partners, Sydney
Pacific Atlas Development Corp., Los Angeles
Peddle Thorp Australia Pty. Ltd., Brisbane
Peddle, Thorp & Walker Arch., Sydney
Perkins & Will, Chicago
J. Roger Preston & Partners, Hong Kong
Projest SA Empreendimentos e Servicos Tecnicos, Rio de Janeiro
Rahulan Zain Associates, Kuala Lumpur
Ranhill Berserkutu Sdn Bhd, Kuala Lumpur
Rankine & Hill, Wellington
RFB Consulting Architects, Johannesburg
Robert Rosenwasser Associates, PC, New York
Emery Roth & Sons Intl., Inc., New York
Rowan Williams Davies & Irwin, Inc., Guelph
Sepakat Setia Perunding (Sdn.) Bhd., Kuala Lumpur
Severud Associates, Cons. Engrs., New York
Shimizu Corporation, Tokyo
SOBRENCO, S.A., Rio de Janeiro
South African Institute of Steel Construction, Johannesburg
Steel Reinforcement Institute of Australia, Sydney
Stigter Clarey & Partners, Sydney
Studio Finzi, Nova E Castellani, Milano
Taylor Thompson Whitting Pty. Ltd., St. Leonards
BA Vavaroutas & Associates, Athens
Pedro Ramirez Vazquez, Arquitecto, Pedregal de San Angel
VIPAC Engineers & Scientists Ltd., Melbourne
Wargon Chapman Partners, Sydney
Weidlinger Associates, New York
Wimberley, Allison, Tong & Goo, Newport Beach
Woodward-Clyde Consultants, New York
Yapi Merkezi Inc., Istanbul
Zaldastani Associates, Inc., Boston

Other Books in the Tall Buildings and Urban Environment Series

Cast-in-Place Concrete in Tall Building Design and Construction
Cladding
Building Design for Handicapped and Aged Persons
Semi-Rigid Connections in Steel Frames
Fatigue Assessment and Ductility Assurance in Tall Building Design
Fire Safety in Tall Buildings

Tall Steel Buildings

Cold-Formed Steel in Tall Buildings

Council on Tall Buildings and Urban Habitat

Committee S37

CONTRIBUTORS

Rolf Baehre
Jan Brekelmans
Byron Daniels
Ken P. Chong
Steven R. Fox
Richard B. Heagler
Larry D. Luttrell
Pieter van der Merwe
Reini M. Schuster
Jan Stark
Derek L. Tarlton
Ton Tomà
Don S. Wolford
Douglas Yates
Wei-Wen Yu

Editorial Group

Wei-Wen Yu Chairman
Rolf Baehre Vice-Chairman
Ton Tomà Editor

McGraw-Hill, Inc.

New York San Francisco Washington, D.C. Auckland Bogotá
Caracas Lisbon London Madrid Mexico City Milan
Montreal New Delhi San Juan Singapore
Sydney Tokyo Toronto

ACKNOWLEDGEMENT OF CONTRIBUTIONS

This Monograph was prepared by Committee S37 (Cold-Formed Steel) of the Council on Tall Buildings and Urban Habitat as part of the *Tall Buildings and Urban Environment Series*. The editorial group was Wei-Wen Yu, chairman; Rolf Baehre, vice-chairman; and Ton Tomà, editor.

Special acknowledgement is due to those who have contributed to Council Report 903.370 (Workshop Proceedings, Third International Conference on Tall Buildings, January 6, 1986), which has served as the basis for this Monograph. These contributors were R. Baehre, K. P. Chong, S. R. Fox, R. B. Heagler, R. M. Schuster, D. L. Tarlton, P.v.d. Merwe, D. S. Wolford, D. Yates, and W. W. Yu.

The individuals whose contributions and papers formed the major contribution to the chapters of this volume are:

Wei-Wen Yu, Chapters 1 and 7
Reini M. Schuster, Section 2.1
Steven R. Fox, Sections 2.1 and 4.2
Derek L. Tarlton, Section 2.1
Don S. Wolford, Section 2.2
Jan Brekelmans, Section 3.1
Byron Daniels, Section 3.1
Richard B. Heagler, Section 3.1
Jan Stark, Section 3.1
Ken P. Chong, Section, 3.2
Larry D. Luttrell, Section 4.1
Douglas Yates, Section 4.2
Ton Tomà, Chapter 5
Rolf Baehre, Chapter 6
Pieter van der Merwe, Chapter 7

COMMITTEE MEMBERS

Rolf Baehre, Ken P. Chong, Steven R. Fox, Richard B. Heagler, A. L. Johnson, K. Klöppel, R. A. LaBoube, Larry D. Luttrell, G. Moreau, Ton Tomà, Pieter van der Merwe, Don S. Wolford, and Wei-Wen Yu.

GROUP LEADERS

The Committee on Cold-Formed Steel is part of Group SB of the Council, "Tall Steel Buildings." The leaders are:

Jerome S. B. Iffland, Group Chairman
Leo Finzi, Group Vice-Chairman
Franklin Y. Cheng, Group Editor
Minoru Wakabayashi, Group Editor
Le-Wu Lu, Group Advisor

Foreword

This volume is one of a new series of Monographs prepared under the aegis of the Council on Tall Buildings and Urban Habitat, a series that is aimed at updating the documentation of the state-of-the-art of the planning, design, construction, and operation of tall buildings and their interaction with the urban environment of which they are a part.

The original Monographs contained 52 major topics collected in the following five volumes:

Volume PC: Planning and Environmental Criteria for Tall Buildings
Volume SC: Tall Building Systems and Concepts
Volume CL: Tall Building Criteria and Loading
Volume SB: Structural Design of Tall Steel Buildings
Volume CB: Structural Design of Tall Concrete and Masonry Buildings

Following the publication of a number of updates to these volumes, it was decided by the Steering Group of the Council to develop a new series that would be based on the original effort but would focus more strongly on the individual topical committees rather than the groups. This would do two things. It would free the coverage of an individual topic from restraints as to length. Also it would permit material on a given topic to more quickly reach the public.

This particular Monograph was prepared by the Council's Committee S37, Cold-Formed Steel. Starting with the solid background provided by the workshop on Cold-Formed Steel held in Chicago in 1986, and the resulting proceedings, the committee went on to supplement and revise that material. This Monograph covers applications to structural members, decking and sandwich panels, shear diaphragms, and connections, as well as special usages in tall buildings and in the application of stainless steel.

THE MONOGRAPH CONCEPT

The Monograph series *Tall Buildings and Urban Environment* is prepared for those who plan, design, construct, or operate tall buildings, and who need the latest information as a basis for judgment decisions. It includes a summary and condensation of research findings for design use, it provides a major reference source to recent literature and to recently developed design concepts, and it identifies needed research.

The Council's Monograph series is not intended to serve as a primer. Its function is to communicate to all knowledgeable persons in the various fields of expertise the state of art and most advanced knowledge in those fields. The message has more to do with setting policies and general approaches than with detailed applications. It aims to provide adequate information for experienced general practitioners confronted with their first high-rise, as well as opening new vistas to those who have been involved with them in the past. It aims at an international scope and interdisciplinary treatment.

The Monograph series was not designed to cover topics that apply to all buildings in general. However, if a subject has application to *all* buildings, but also is particularly important for a *tall* building, then the objective has been to treat that topic.

Direct contributions to this Monograph have come from many sources. Much of the material has been prepared by those in actual practice as well as by those in the academic sector. The Council has seen considerable benefit accrue from the mix of professions, and this is no less true in the Monograph series itself.

TALL BUILDINGS

A tall building is not defined by its height or number of stories. The important criterion is whether or not the design is influenced by some aspect of "tallness." It is a building in which "tallness" strongly influences planning, design, construction, and use. It is a building whose height creates conditions different from those that exist in "common" buildings of a certain region and period.

THE COUNCIL

The Council is an international group sponsored by engineering, architectural, construction, and planning professionals throughout the world, an organization that was established to study and report on all aspects of the planning, design, construction, and operation of tall buildings.

The sponsoring societies of the Council are the American Institute of Architects (AIA), American Society of Civil Engineers (ASCE), the American Planning Association (APA), American Society of Interior Designers (ASID), International Association for Bridge and Structural Engineering (IABSE), International Federation of Interior Designers (IFI), International Union of Architects (UIA), Japan Structural Consultants Association (JSCA), and the Urban Land Institute (ULI).

The Council is concerned not only with the building itself but also with the role of tall buildings in the urban environment and their impact thereon. Such a concern also involves a systematic study of the whole problem of providing adequate space for life and work, considering not only technological factors, but social and cultural aspects as well.

The Council is not an advocate for tall buildings per se; but in those situations in which they are viable, it seeks to encourage the use of the latest knowledge in their implementation.

NOMENCLATURE AND OTHER DETAILS

The general guideline for units is to use SI units first, followed by American units in parentheses, and also metric when necessary. A conversion table for units is supplied at the end of the volume. A glossary of terms and a list of symbols also appears at the end of the volume.

The spelling was agreed at the outset to be "American" English.

A condensation of the relevant references and bibliography will be found at the end of each chapter. Full citations are given in a composite list at the end of the volume.

From the start, the Tall Building Monograph series has been the prime focus of the Council's activity, and it is intended that its periodic revision and the implementation of its ideas and recommendations should be a continuing activity on an international level. Readers who find that a particular topic needs further treatment are invited to bring it to our attention.

ACKNOWLEDGMENTS

This work would not have been possible but for the early financial support of the National Science Foundation which supported the program out of which this Monograph developed. More recently the major financial support has been from the organizational members, identified in earlier pages of this Monograph, as well as from many individual members. Their confidence is appreciated. Specific support of individual chapters has been recognized in the Preface.

All those who had a role in the authorship of the volume are identified in the acknowledgment page that follows the title page. Especially important are the contributors whose papers formed the essential first drafts—the starting point.

The primary conceptual and editing work was in the hands of the leaders of the Council's Committee S37, Cold-Formed Steel. The Chairman of this editorial group is Wei-Wen Yu of University of Missouri, Rolla, USA, who also contributed substantially to the editing. Vice-Chairman is Rolf Baehre of Universitat Karlsruhe, Karlsruhe, Germany. Ton Tomà of TNO Building and Construction Research, Rijswijk, The Netherlands is the editor.

Overall guidance was provided by the Group Leaders Jerome S. B. Iffland of Iffland Kavanaugh Waterbury, New York, USA; Prof. Dr. Ing. Leo Finzi of Politechnico di Milano, Milano, Italy; Mr. Franklin Y. Cheng of University of Missouri, Rolla, USA; and Prof. Minoru Wakabayashi of General Building Research Corp., Suita City, Japan. Le-Wu Lu of Lehigh University served as Group Advisor.

Lynn S. Beedle
Editor-in-Chief

Dolores B. Rice
Managing Editor

Lehigh University
Bethlehem, Pennsylvania
1993

Preface

In high-rise steel buildings, cold-formed steel is widely used in floor and wall construction. This Monograph on Cold-Formed Steel was prepared to provide readers with information needed for the design and construction of tall buildings, using cold-formed steel for structural members and/or architectural components.

This Monograph contains seven chapters on applications and various design features of cold-formed steel members and connections fabricated from carbon, low-alloy, and stainless steel sheet, strip and plate. Fifteen contributors were involved, and even though an attempt has been made to achieve a complete international spectrum of development, the emphasis is on Europe, North America, and South Africa. Readers are invited to offer their suggestions and further contributions for future enhancement of the Monograph.

In the 1986 Workshop on Cold-Formed Steel, held during the Third International Conference on Tall Buildings in Chicago, seven specific presentations and discussions were concerned with the following subjects on cold-formed steel:

"Steel Joists," by D. S. Wolford
"A LRFD Standard for Cold-Formed Steel Design," by R. M. Schuster, S. R. Fox, and D. L. Tarlton
"Composite Floor Deck in Tall Buildings," by R. E. Heagler
"Sandwich Panels," by K. P. Chong
"A Design Method for Steel Deck Diaphragms," by S. R. Fox and D. Yates
"Stainless and Corrosion Resisting Steel Sections," by P. van der Merwe
"Functional Requirements for Tall Buildings," by R. Baehre

These manuscripts were published in the Council's 1986 Workshop proceedings (Council on Tall Buildings, 1987), and have been used as basic material for drafting this present Monograph. This volume can be used not only by engineers and architects but also by specification writers to improve the use and design of cold-formed steel for tall buildings.

Wei-Wen Yu
Chairman

Rolf Baehre
Vice-Chairman

Ton Tomà
Editor

Contents

Cold-Formed Steel in Tall Buildings

1 Introduction

Typical cold-formed steel structural members, as shown in Figs. 1.1 and 1.2, are composed of sections cold-formed to shape from steel sheet, strip, or plate by means of roll-forming machines or press brake operations. The thicknesses of the materials most frequently used for this type of structural members range from 0.4 mm (0.015 in.) to about 6.4 mm (0.25 in.) (Yu, 1991). Today thicknesses of up to 12 mm ($^1/_2$ in.) can be cold-formed to sections.

Cold-formed steel members have been used for buildings since 1850. From a structural viewpoint, these types of steel members can be classified as (1) individual structural framing members and (2) decks and panels. In view of the fact that light members of various configurations can be easily cold-formed to structural shapes for carrying light loads, such sections have been used successfully in numerous high-rise steel buildings to supplement hot-rolled steel shapes.

The purposes of this chapter are to discuss the applications of cold-formed steel members in tall buildings, to review some of the available design standards used for cold-formed steel structural components and structural systems, and finally to touch on some functional requirements.

1.1 APPLICATIONS OF COLD-FORMED STEEL

Cold-formed steel sections can be used as primary and secondary framing members in low-rise buildings up to eight stories in height. For high-rise steel buildings, the main frame is typically composed of heavy, hot-rolled structural steel shapes. Cold-formed steel members, such as roof and floor decks, steel joists, wall panels, sandwich panels, partitions, door and window frames, duct systems, and entrance structures, have been successfully used for tall buildings in a supplementary and complementary manner (Yu, 1991; SDI, 1987; Iyengar and Zils, 1973; Architectural Record, 1976). Recent applications also include the construction of unitized boxes. These boxes are prefabricated at room size, fully furnished, and stacked in some manner to form structures for hotels, apartments, and office buildings. For multistory buildings, these boxes can be supported by a main frame made of hot-rolled steel. These boxes can also be applied as wet or service cells in buildings of other structural types (see IISI, 1988).

In building construction, steel roof and floor decks are used in either composite or non composite systems (SDI, 1987), as discussed in Chapter 3. The economical use of cold-formed steel decks in composite floor construction has been accelerated by numerous studies and the issuance of various design specifications and related publications (Ekberg and Schuster, 1968; Canadian Steel Sheet Building Institute, 1968; Robinson, 1969; Fisher, 1970; Luttrell and Davison, 1973; Iyengar and Zils, 1973; Badoux and Crisinel, 1973; ECCS, 1975; Porter and Ekberg, 1976; Grant et al., 1977; Stark, 1978; Temple and Abdel-Sayed, 1978; ECCS, 1981; Bucheli and Crisinel, 1982; Porter and Greimann, 1982; Seleim and Schuster, 1982; Porter and Greimann, 1984; ASCE, 1984; Luttrell, 1986; Porter, 1986; Schurter and Schuster, 1986; Porter, 1988; Bode et al., 1988; McCraig and Schuster, 1988; Easterling and Porter, 1988; AISC, 1989; Brekelmans et al., 1990; Daniels, 1990; Eurocode 4 Part 1, 1990; Stark and Brekelmans, 1990). Table 1.1 lists some of the tall buildings using composite steel floor decks as structural components of floor slabs (Albrecht, 1989; Heagler, 1989). Based on the Dodge Reports, the estimated annual volume of composite steel deck slab construction for tall buildings exceeds 9 million m^2 (100 million ft^2) in the United States (Albrecht, 1989). Using an approximate weight for steel decks of 96 Pa (2 psf), more than 100,000 tons of sheet steel is used in the United States every year for composite slab construction for tall buildings.

Usually the depths of steel decks range from 38 to 191 mm (1.5 to 7.5 in.) and the thicknesses of the materials vary from about 0.5 to 1.9 mm (0.018 to 0.075 in.). The availability of different profiles has resulted in economical designs for a wide range of building projects.

From the viewpoint of the designer, steel decks not only provide structural strength to carry normal loads, they also provide surfaces for roofs, floors, or concrete fill. In addition, such structural elements can provide spaces for electrical conduits or can be perforated and combined with sound-absorption material to form an acoustically conditioned ceiling. The cells of cellular panels are also often used as ducts for heating and air-conditioning (see also Chapter 6 and ECCS, 1990a).

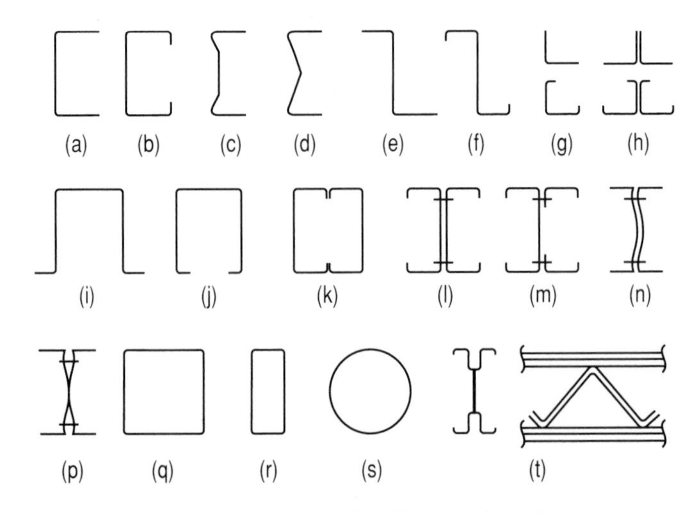

Fig. 1.1 Typical cold-formed steel structural members.

In the past, dry floor systems in which steel decks are used with plywood, gypsum board, and mineral wool have also been developed for tall buildings (Baehre and Urschel, 1984). The effects of composite action in bending on the load-deformation relationships for composite beams using several materials have been demonstrated by Baehre et al. (1982). In addition, steel joists have been used widely for a large number of dry floor systems (Newman, 1966; Wang and Kaley, 1967; Tide and Galambos, 1970; Cran, 1971), and prefabricated truss-

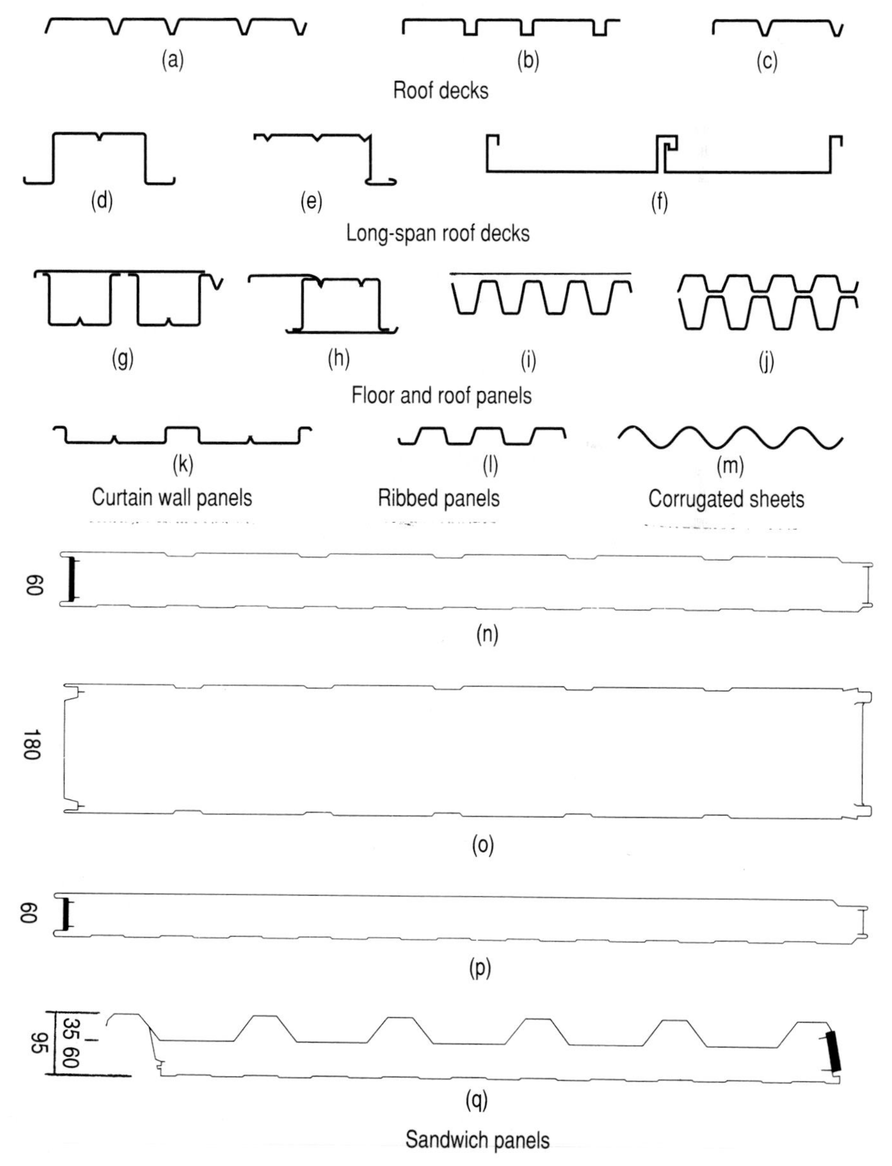

Fig. 1.2 Decks, panels, corrugated sheets, and sandwich panels.

panel systems have been used successfully in floor construction such as the World Trade Center in New York City (Yu, 1991).

In high-rise building construction, cold-formed steel studs are used widely for exterior nonbearing curtain walls, interior drywall partitions, and cavity shaft walls (see Council on Tall Buildings, Group SC, 1980, chapter 6; Council on Tall Buildings, 1983, chapter SC-5). All the systems use gypsum board and other panel materials attaching to the studs. Virtually all buildings over four stories use drywall steel studs and cavity shaft walls (Marchello, 1989).

In recent years some of the roof decks, floor decks, and wall panels have been designed as shear diaphragms because they are capable of resisting horizontal loads in their own planes. These diaphragms also have beam strength for carrying normal loads as long as the structural components are adequately interconnected to each other and to the supporting frame. The problem of controlling drift through the use of cold-formed steel infill panels has also been studied (Miller, 1973, 1974; Tomasetti et al., 1986).

1.2 DESIGN STANDARDS AND RECOMMENDATIONS

1 Design Specifications and Standards for Structural Members

At the present time, specifications and standards for the design of cold-formed steel structural members using sheet steels are available in numerous countries. A

Table 1.1 Tall buildings (above 40 stories in height) using composite steel floor decks

Building and location	Height, stories	Depth of composite steel deck, in.
Sears Tower, Chicago, Ill., USA	110	3
World Trade Center, New York, N.Y., USA	104	$1^1/_4$
First Interstate World Center, Los Angeles, Calif., USA	73	3
Columbia Building (office building), Seattle, Wash., USA	70	3
USX Building, Pittsburgh, Pa., USA	63	3
Bank of America, San Francisco, Calif., USA	60	3
Transamerica Building, San Francisco, Calif., USA	54	3
Bell Atlantic Building, Philadelphia, Pa., USA	52	3
Treasury Building, Singapore	50	3
Americas Tower, New York, N.Y., USA	48	3
Financial Square, New York, N.Y., USA	47	3
599 Lexington Ave. (office building), New York, N.Y., USA	47	3
International Place, Boston, Mass., USA	46	3
Hong Kong and Shanghai Bank, Hong Kong	44	$2^1/_8$

1 in. = 25.4 mm.

brief comparison of several design specifications has been made by Beedle (1991, chap. 11). Insofar as the design method is concerned, some specifications use the allowable stress design (ASD) approach, whereas others are based on limit state design (LSD), or load and resistance factor design (LRFD). It is likely that in future editions of the specifications of the American Iron and Steel Institute (AISI) both ASD and LRFD methods will be included for the design of cold-formed steel structural members and connections (AISI, 1986, 1991a, 1991b; Hsiao et al., 1990). In Europe the Eurocodes are all based on limit states. As far as thin-walled steel is concerned, Eurocode 3 Annex A (1991) and Eurocode 4 Part 1 (1990) are relevant. In the framework of the European Convention for Constructional Steelwork (ECCS) recommendations have also been drafted (ECCS, 1983a, 1983b, 1987). In the United Kingdom the British standard BS 5950: Part 5 (1987) and in Germany the DASt-Richtlinie 016 (1988) treat the relevant subjects.

In the United States, in addition to the design criteria for carbon and low-alloy sheet steels, AISI (1974) also published a specification for the design of stainless-steel structural members. The American Society of Civil Engineers' (ASCE) standard for stainless-steel design has been developed recently (Lin et al., 1988; ASCE, 1991). As far as sandwich panels are concerned, in Europe two recommendations have been drafted by ECCS Committee TC7 for Sandwich Panels (ECCS, 1990b, 1991).

2 Structural Systems

Composite Slabs and Composite Beams with Cold-Formed Steel Deck. As discussed, cold-formed steel decks have been used successfully in the construction of composite roofs and floors. In this type of application, a steel deck performs a dual role. It serves as a form to support wet concrete during construction and as positive reinforcement for the slab during service. In the United States, many designs of this type of composite construction are based on a Steel Deck Institute (SDI) specification (SDI, 1987). Recently a specification for the design and construction of composite steel deck slabs has been developed by ASCE's Technical Council on Codes and Standards (ASCE, 1984). These criteria are based mainly on the research work conducted at Iowa State University under the sponsorship of AISI (Ekberg and Schuster, 1968; Porter and Ekberg, 1976; ASCE, 1984).

Since 1978, the specifications of the American Institute of Steel Construction (AISC) have included specific provisions for the design of composite beams with cold-formed steel decks (AISC, 1989). These provisions are based primarily on the research work conducted at Lehigh University (Fisher, 1970; Grant et al., 1977).

In other countries the design and use of composite construction have been discussed by Robinson (1969), Badoux and Crisinel (1973), Stark (1978), Bucheli and Crisinel (1982), Bode et al. (1988), Brekelmans et al. (1990), Daniels (1990), and Stark and Brekelmans (1990). All this work had lead to European recommendations (ECCS, 1975, 1981) and finally to Eurocode 4 Part 1 (1990).

Shear Diaphragms. Steel decks and panels have been used as shear diaphragms subsequent to extensive studies conducted by a large number of investigators in North America and Europe. In designing shear diaphragms, consideration should be given to the shear strength and stiffness of the system. This can be done by using either tests or analytical methods. Based on the available information, in the past the following methods have been used for shear diaphragms (Yu, 1991):

1. AISI method: design of shear diaphragms (AISI, 1967)
2. Stressed skin design of steel buildings (Bryan, 1972)
3. Tri-Services method (TRI, 1973)
4. European recommendations for the stressed skin design of steel structures (ECCS, 1977)
5. Simplified diaphragm analysis (Davies, 1977)
6. SDI method (Luttrell, 1980)
7. Nonlinear finite-element analysis (Atrek and Nilson, 1980)

With regard to the use of European recommendations, design guides were published (Bryan and Davies, 1981, 1982). These publications contain useful design information for calculating the strength and stiffness of steel roof decks when acting as diaphragms. For additional information on shear diaphragms in high-rise buildings, see Council on Tall Buildings, Group SC (1980).

1.3 FUNCTIONAL REQUIREMENTS

In addition to the design requirements for strength and stiffness, consideration should also be given to some functional requirements such as fire protection, noise control, vibrational characteristics, thermal insulation, and corrosion protection. These requirements were discussed by Baehre and Urschel (1984). ECCS (1984) has also published some documents concerning this item. Additional information can also be found in other references (Newman, 1966; Wang and Kaley, 1967). The static and dynamic responses of cold-formed steel joist residential floor systems were discussed by Kudder et al. (1978) and Linehan et al. (1978).

1.4 CONDENSED REFERENCES/BIBLIOGRAPHY

AISC 1989, *Specification for the Design, Fabrication and Erection of Structural Steel for Buildings*

AISI 1967, *Design of Shear Diaphragms*

AISI 1974, *Specification for the Design of Cold-Formed Stainless Steel Structural Members*

AISI 1986, *Specification for the Design of Cold-Formed Steel Structural Members*

AISI 1991a, *Load and Resistance Factor Design Specification for Cold-Formed Steel Structural Members*

AISI 1991b, *Commentary on the Load and Resistance Factor Design Specification for Cold-Formed Steel Structural Members*

Albrecht 1989, *Use of Cold-Formed Steel as Composite Floor Deck in Tall Buildings*

Architectural Record 1976, *Light-Gage Steel in the Framing for Lightweight Wall Panels*

ASCE 1984, *Specification for the Design and Construction of Composite Slabs and Commentary on Specifications for the Design and Construction of Composite Slabs*

ASCE 1991, *Specification for the Design of Cold-Formed Stainless Steel Structural Members, ANSI/ASCE-8-90*

Atrek 1980, *Nonlinear Analysis of Cold-Formed Steel Shear Diaphragms*

Badoux 1973, *Recommendations for the Application of Cold-Formed Steel Decking for Composite Slabs in Buildings*

Baehre 1982, *Cold-Formed Steel Applications Abroad*

Baehre 1984, *Light-Weight Steel Based Floor Systems for Multi-Story Buildings*

Beedle 1991, *Stability of Metal Structures, A World View, 2d ed.*

Bode 1988, *Profiled Steel Sheeting and Composite Action*

Brekelmans 1990, *Comparative Study of Composite Slab Tests*

Bryan 1972, *The Stressed Skin Design of Steel Buildings*

Bryan 1981, *Steel Diaphragm Roof Decks, A Design Guide with Tables for Engineers and Architects*

Bryan 1982, *Manual of Stressed Skin Diaphragm Design*

BS 5950: Part 5 1987, *Structural Use of Steelwork in Building, Part 5: Code of Practice for Design of Cold Formed Sections*

Bucheli 1982, *Composite Beams in Buildings*

Canadian Sheet Steel Building Institute 1968, *Composite Beam Manual for the Design of Steel Beams with Concrete Slab and Cellular Steel Floor*

Council on Tall Buildings 1983, *Developments in Tall Buildings 1983*

Council on Tall Buildings Group SC 1980, *Tall Building Systems and Concepts*

Cran 1971, *Design and Testing Composite Open Web Steel Joists*

Daniels 1990, *Bearing Capacity of Composite Slabs: Mathematical Modeling and Experimental Study*

DASt-Richtlinie 016 1988, *Calculation and Design of Structures from Cold-Formed Building Elements*

Davies 1977, *Simplified Diaphragm Analysis*

Easterling 1988, *Composite Diaphragm Behavior and Strength*

ECCS 1975, *European Recommendations for the Calculation and Design of Composite Floors with Profiled Steel Sheet*

ECCS 1977, *European Recommendations for the Stressed Skin Design of Steel Structures*

ECCS 1981, *Composite Structures, Model Code*

ECCS 1983a, *European Recommendations for the Design of Profiled Sheeting*

ECCS 1983b, *European Recommendations for Good Practice in Steel Cladding and Decking*

ECCS 1984, *Lightweight Steel Based Floor Systems for Multi-Storey Buildings*

ECCS 1987, *European Recommendations for the Design of Light Gauge Steel Members*

ECCS 1990a, *European Recommendations for Sound Insulation of Steel Construction in Multi-Storey Buildings*

ECCS 1990b, *European Recommendations for Sandwich Panels, Part II: Good Practice*

ECCS 1991, *European Recommendations for Sandwich Panels, Part I: Design*

Ekberg 1968, *Floor Systems with Composite Form-Reinforced Concrete Slabs*

Eurocode 3 Annex A 1991, *Cold Formed Steel Sheeting and Members*

Eurocode 4 Part 1 1990, *Design of Composite Steel and Concrete Structures, Part 1—General Rules and Rules for Buildings*

Fisher 1970, *Design of Composite Beams with Formed Metal Deck*

Grant 1977, *Composite Beams with Formed Steel Deck*

Heagler 1989, *Composite Floor Deck*

Hsiao 1990, *AISI LRFD Method for Cold-Formed Steel Structural Members*

IISI 1988, *Steel in Housing*

Iyengar 1973, *Composite Floor Systems for Sears Tower*

Kudder 1978, *Static and Ultimate Load Behavior of Cold-Formed Steel-Joists Residential Floor Systems*

Lin 1988, *ASCE Standard for Stainless Steel Structures*

Linehan 1978, *Dynamic and Human Response Behavior of Cold-Formed Steel-Joists Residential Floor Systems*

Luttrell 1973, *Composite Slabs with Steel Deck Panels*

Luttrell 1980, *Steel Deck Institute Diaphragm Design Manual*

Luttrell 1986, *Methods for Predicting Strength in Composite Slabs*

Marchello 1989, *Steel-Framed Wall Applications*

McCraig 1988, *Repeated Point Loading on Composite Slabs*

Miller 1973, *Drift Control with Light Gage Steel Infill Panels*

Miller 1974, *Light Gage Steel Infill Panels in Multistory Steel Frames*

Newman 1966, *The Dry Floor—A New Approach to High Rise Apartment Buildings*

Porter 1976, *Design Recommendations for Steel Deck Floor Slabs*

Porter 1982, *Composite Steel Deck Diaphragm Slabs—Design Modes*

Porter 1984, *Shear-Bond Strength of Studded Steel Deck Slabs*

Porter 1986, *Highlights of New ASCE Standard on Composite Slabs*

Porter 1988, *Two-Way Analysis of Steel Deck Floor Slabs*

Robinson 1969, *Composite Beam Incorporating Cellular Steel Decking*

Schurter 1986, *Aluminium–Zinc Alloy Coated Steel for Composite Slabs*

SDI 1987, *Design Manual for Composite Decks, Form Decks and Roof Decks*

Seleim 1982, *Shear-Bond Capacity of Composite Slabs*

Stark 1978, *Design of Composite Floors with Profiled Steel Sheet*

Stark 1990, *Plastic Design of Continuous Slabs*

Temple 1978, *Fatigue Experiments on Composite Slab Floors*

Tide 1970, *Composite Open-Web Steel Joists*

Tomasetti 1986, *Development of Thin Wall Cladding to Reduce Drift in Hi-Rise Buildings*

TRI 1973, *Seismic Design for Buildings*

Wang 1967, *Composite Action of Concrete Slab and Open Web Joist (Without the Use of Shear Connectors)*

Yu 1991, *Cold-Formed Steel Design*

2

Structural Members

2.1 LRFD STANDARD FOR COLD-FORMED STEEL DESIGN

In December 1984 the Canadian Standards Association (CSA) published a new design standard entitled *Cold Formed Steel Structural Members* (CAN3-S136-M84, 1984). It superseded the decade-old previous edition with the same title, but designated S136-1974. Publication of the 1984 edition culminated more than 6 years of work by the CSA technical committee responsible for the contents. [This edition was since revised in 1989 to incorporate additional technical advances, in keeping with the 5-year revision cycle of the National Building Code of Canada (NBCC). In the 1989 standard (CAN3-S136-M89, 1989), the major change is a "unified" effective-width approach for the design of compressive elements subject to local buckling. The treatment of compressive elements with intermediate and edge stiffeners has also been revised to allow for the partially stiffened case.]

In contrast to the 1974 edition, which was based on allowable stress design with a limit state design option, the 1984 standard is based entirely on limit state design principles, also referred to as load and resistance factor design (LRFD). Although the standard has been prepared for use with SI (metric) units, the designer is able to substitute any other consistent units of measurement by using the applicable general expressions provided alongside the SI expressions. In the United States (AISI, 1991a, 1991b) and in Europe (Eurocode 3 Annex A, 1991) also, LRFD specifications are used for cold-formed steel design.

1 An LRFD Approach

Compared with allowable stress design, LRFD affords a better understanding of the relationship between the performance requirements of a member and its behavior under loads at the limit of structural usefulness as well as its performance under the smaller loads anticipated in service. LRFD also permits the adoption of a common format in design standards for various materials and in the codes governing end use, such as building codes. The limit state format separates the code parameters, accounting for the uncertainties associated with the determination of loads, from the design standard parameters, accounting for the uncertainties associated with the determination of member resistances. The basic format is shown in Table 2.1.

Probabilistic methods provide a means of calibrating the uncertainties associated with the determination of both loads and resistances, and some of the more germane literature on the subject is provided in Allen (1975, 1981), CSA S408-1981 (1981), Kennedy and Gad Aly (1980), Nowak and Lind (1979), and Parimi and Lind (1976). In brief, the LRFD approach to cold-formed steel design adopted by the technical committee responsible for CAN3-S136-M84 (1984) is given in the following.

2 Calibrating for Structural Reliability in Limit State Design

Traditionally the structural designer has followed the requirements of building codes and design standards to effect a "safe" design without the need to estimate the probability of failure. Building codes stipulate minimum requirements arrived at by a group of persons considered to have expertise in that field. These codes change periodically to reflect the current level of knowledge, and in response to political and economic pressures and the expectations of society. The responsibility placed on the codes and standards writing committees to provide a safe design tends to promote a conservative attitude in the development process, and a desire to maintain a target level of safety.

The real measure of a safe design method is reflected by the failure rate of actual building structures. Acceptable failure rates for different building types and occupancies correspond to a tradeoff between the cost of a failure, including reconstruction costs as well as the intangible costs associated with public safety, and the costs associated with increasing the structural safety margin. In practice, satisfactory reliability levels are facilitated by reference to past experience, competent engineering, and the control of human errors.

Human error is the root cause of most structural failures, and typically these human errors are "gross" errors (Fox, 1983). Gross errors are unpredictable and are not covered by the overall safety factor inherent in buildings. To try to protect against gross errors through safety factors would be prohibitively expensive and unjustifiable in terms of the marginal costs incurred. As a result, building codes and design standards assume that workmanship, material properties, dimensions, and such can be held within a predictable range, and the consequences of gross errors are not specifically taken into account.

In limit state design, structural reliability can be specified in terms of a safety index β, which is determined in the following manner. Let us first define the

Table 2.1 Limit state format

Condition	Material design standard*	Building code (use code)†
Ultimate limit states	Factored resistance ≥	Effect of factored load
Serviceability limit states	Serviceability limit ≥	Effect of specified loads (unfactored)

* Material design standard is unique to material being considered.
† Use code applies to all materials which may be used.

following variables:

R = specified nominal value of resistance
$\bar{R}$ = mean value of actual resistance
S = specified nominal value of load effects
$\bar{S}$ = mean value of actual load effects
Φ = resistance factor
α = aggregate load factor

Figure 2.1 shows possible frequency distributions for both load effects and structural member resistance. The figure presumes that (1) the nominal or assumed value of resistance R is less than the mean actual resistance $\bar{R}$, (2) the nominal or assumed effects of the loads S are greater than the mean actual effects of the loads $\bar{S}$, and (3) the factored resistance ΦR is slightly greater than the effects of the factored loads αS, indicating that the code safety criterion has been met. The possibility of loads being greater than the resistance shows that there is a chance of failure. Such a possibility can be reduced to any small value desired, but at a corresponding expense.

The data on actual loads and resistances indicate that a logarithmic transformation is warranted to transform the data into a more "normal" distribution. Figure 2.2 shows the frequency distribution of the algebraic difference between the loads and the resistance using log values, $u = \ln R - \ln S$. The area of the curve that is to the left of zero, $u < 0$, is proportional to the probability of failure. The distribution shown in Fig. 2.2 has a mean value approximately equal to $\bar{u} \simeq \ln \bar{R} - \ln \bar{S}$, and the mean is a certain number of standard deviations from the zero point. The safety index β is defined as the number of standard deviations the mean of this distribution is from zero.

The safety index is directly related to the structural reliability of the design. Increasing β will increase the reliability, and decreasing β decreases the reliability. β is also directly related to the load and resistance factors used in the design.

The NBCC (1985), with the introduction of limit state design, has defined a set of load factors, load combination factors, and a set of specified minimum loads to

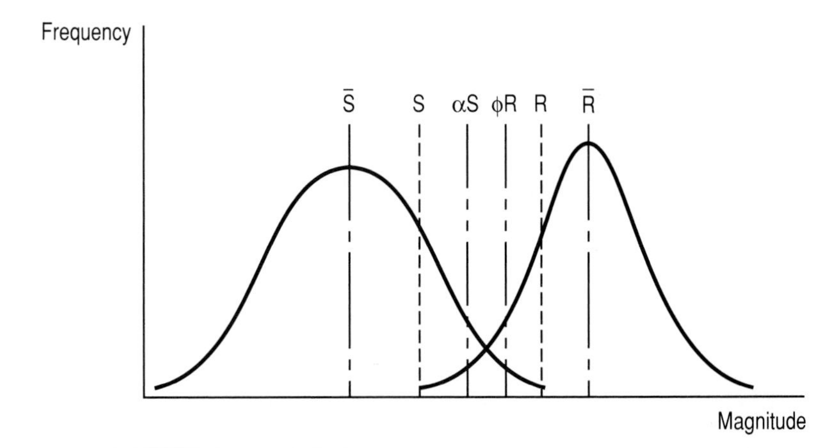

Fig. 2.1 Frequency distributions for load effects and structural member resistance.

be used in the design. The specification of these loads and load factors has fixed the position of the nominal load distribution S and the factored load distribution αS. The material design standard is then obligated to specify the appropriate resistance function.

Those responsible for writing a design standard are given the loading distribution and load factors and must calibrate the resistance factors Φ such that the safety index β reaches a certain target value. The technical committee responsible for the S136 standard elected to use a target safety index equal to 3.5, in keeping with a similar level used for other structural materials.

To satisfy the basic safety criterion, all factored resistances must be equal to or greater than the effect of the factored loads. The factored resistance of any given member is given by the product ΦR, with Φ being the resistance factor and R the nominal or theoretical member strength. The resistance factors specified for strength analysis in the 1984 standard are given in Table 2.2.

3 Summary of Specification Changes (1974 to 1984)

Since the structural design of cold-formed steel members is a relatively modern development based on extensive testing, there exist a great deal of test data and documentation to assist in the derivation of appropriate resistance expressions for various limit states (tension, compression, bending, shear, and such). Further examination of these test data in comparison with calculated values provided evidence that modification of the 1974 requirements for axial compression, combined axial compression and bending, effective widths, and bolted connections was desirable. Also, research developments had shown that improvements were possible in the specification of requirements for web bending, web crippling, welding, and screw-type fasteners. Thus a number of changes have been incorporated in these areas as well. All of the changes reflect an increased understanding of the behavior of cold-formed steel structures, members, and elements, and of cold-formed steel as a structural material. The more significant aspects of the new LRFD standard issued by CSA and some comparisons with provisions of its predecessor, and also, where appropriate, with the 1980 AISI specification (AISI, 1980), are summarized as follows.

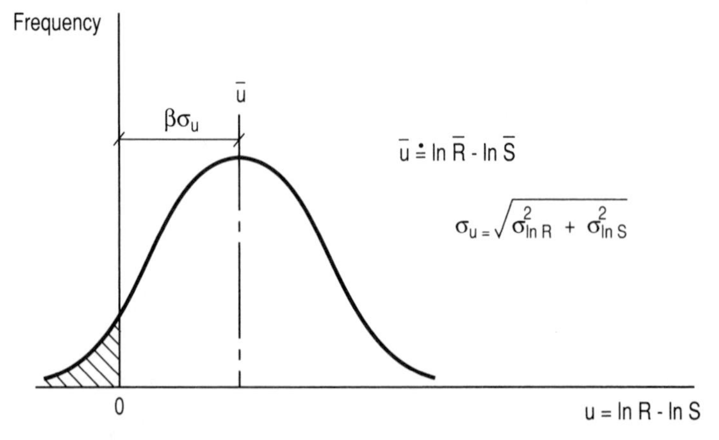

Fig. 2.2 Frequency distribution of algebraic difference between loads and resistance.

Local Buckling and Properties of Sections. Since local buckling of cold-formed steel elements subjected to compression in flexural bending, axial compression, shear, or bearing can occur at stress levels below the yield strength of the steel, postbuckling becomes an important consideration in the design of cold-formed steel members. The well-known phenomenon of postbuckling in thin compressed plates is reflected in the effective-width approach used in both Canada and the United States when computing section properties of stiffened compression elements. It has been long-standing practice in both countries to compute section properties of stiffened compression elements on the basis of an effective-width concept, in other words, reduced section properties and full allowable stress in comparison with a reduced stress on the gross or full section as used in the design of unstiffened compression elements.

CAN3-S136-M84 (1984) uses an effective-width (reduced section properties) approach for both stiffened and unstiffened compression elements, thus providing the designer with a more consistent method.

Members in Tension. Significant changes have been made in the 1984 S136 standard in comparison with the 1974 edition and the 1980 AISI specification with respect to members in tension.

The ultimate limit state is defined as that condition in which the member suffers either uncontrolled deformation or collapse. Collapse may be caused by either rupture or instability of the metal. Uncontrolled deformation is deemed to occur when the gross section yields over a significant length. This gives a limiting state which, while rendering the structure useless, gives some warning of impending failure and may not be the true highest load capacity. The factored resistance for this condition is given by

$$T_r = \Phi A_g F_y \tag{2.1}$$

Should yielding across the net section be exceeded, it is confined to a narrow band, and the absolute overall elongation of the member is comparable to the elastic extension. Thus yielding in this zone does not lead to an ultimate limit state. Failure occurs when the stress at the net section reaches the ultimate value. As this condition is not preceded by any gross deformation and failure is precipitous, the resistance factor is lower than that for overall yielding. The factored resistance for this condition is given by

$$T_r = \Phi_u A_n F_u \tag{2.2}$$

Table 2.2 Resistance factors for strength analysis (*CAN3-S136-M84, 1984*)

Type of strength	*Resistance factor*
Axial tension, bending, and shear	$\Phi = 0.90$
Axial compression:	
Doubly symmetric sections, angles, and wall studs	$\Phi_a = 0.90$
Bearing stiffeners and other sections	$\Phi_a = 0.75$
Web crippling in beams:	
Single unreinforced webs and deck sections	$\Phi_s = 0.80$
Other webs	$\Phi_o = 0.67$
Connections	$\Phi_c = 0.67$
Limit states determined by tensile strength of material	$\Phi_u = 0.75$

The lesser of these two values will control the design. This design procedure clearly differentiates between the two types of ultimate states, recognizing the different relationships between yield and ultimate stresses without attempting to qualify one by the other.

Members in Bending. The 1984 S136 standard uses the same philosophical approach as AISI (1980) to govern against lateral-torsional buckling of symmetrical I-shaped beams or channels and of point-symmetrical Z-shaped sections. The only difference is that the CSA standard expresses the governing equations in a resistance-type format, namely, the factored moment resistance of a member in bending equals

$$M_r = \Phi SF \tag{2.3}$$

where Φ is either 0.90 or 0.75, depending on the failure mode considered, S is the section modulus for either compression or tension for four different cases, and F is a stress function that depends on either the yield strength of the steel F_y, the tensile strength of the steel F_u, the lateral-torsional buckling stress F_c, or the web bending stress F_{wb}, whichever results in the lesser moment resistance.

Bending in Webs. The structural behavior of cold-formed steel beam webs subjected to pure bending has been studied in detail, and it was found that the postbuckling resistance of web elements under pure bending is a function of four significant parameters: the web slenderness ratio H, the tension-to-compression stress ratio β, the flat-width ratio of the flange W, and the yield strength of the material F_y. Based on the results of numerous tests carried out by LaBoube and Yu (1978a), a number of different design methods have been proposed. The 1980 edition of AISI specifies the moment resistance method, which is a simplified approach and does not include the tension-to-compression stress ratio β and the flat-width ratio of the flange W. From an in-depth study of these parameters it was found that the postbuckling resistance increases as H and F_y increase, while β decreases. An increase of W will result in a reduction of the postbuckling resistance.

Shear in Webs. In the 1974 edition of S136, the design expressions for determining the limiting shear stress were developed for beam webs without stiffeners (CAN3-S136-1974, 1974). In the 1984 edition of the standard, however, shear resistance expressions are provided for webs with and without stiffeners. These provisions are based on the results of a study of beam webs loaded primarily by shear stress (LaBoube and Yu, 1978b), and they are also used in the 1980 AISI specification.

Combined Bending and Shear in Webs. This part of the standard provides for the interaction between bending and shear in webs and their effect on the capacity of the web element. The interaction expression from the previous edition of the standard has been included for unreinforced flat webs. In addition, a new interaction equation has been included for use with beam webs with adequate transverse stiffeners (Phung and Yu, 1978), which is also used in AISI (1980).

Web Crippling. Considerable changes have been made in the web crippling expressions in comparison to the 1974 edition of S136. These changes were primarily based on tests carried out in both the United States (Hetrakul and Yu, 1978) and Canada (Wing, 1981). The most significant change occurred in the addition of expressions for two-flange loading, which did not exist in the 1974 edition of S136. These expressions, presented in limit state format in the 1984 standard, were adopted directly from the 1980 AISI specification.

In addition, expressions for the design of deck sections (multiple webs) have been added to the 1984 standard which are not contained in the 1980 AISI specification and did not exist in the 1974 edition of S136. All new web crippling expressions are based solely on testing, and the limits generally placed on the various parameters have been expanded to reflect the findings of the most recent research on web crippling.

Inelastic Reserve Resistance of Members in Bending. The inelastic reserve resistance of laterally supported flexural members is the additional moment which many flexural members develop over and above the yield moment, before the ultimate failure moment is reached; in other words, the inelastic reserve capacity is $M_{ult} - M_{yield}$. This is a new section in the 1984 edition of S136 and was taken directly from the 1980 AISI specification but expressed in an LRFD resistance-type format.

Stiffeners for Beam Webs. This new section has been adopted from the 1980 AISI specification and provides design requirements for bearing and intermediate stiffeners. In the case of bearing stiffeners, the resistance of the transverse stiffeners to end crushing is being considered as well as the resistance of column-type buckling of the web stiffeners. The new expressions for determining the minimum required moment of inertia and the minimum required gross area of attached intermediate stiffeners are based on the studies summarized in Phung and Yu (1978).

Concentrically Loaded Compression Members. AISI has traditionally used the tangent-modulus approach with a constant factor of safety (1.92) in the design of cold-formed steel compression members. The overall column strength is reduced by the introduction of a local buckling factor that reflects the interaction of local and overall buckling in the inelastic region only (defined by a Johnson parabola). The 1974 edition of S136 used a similar approach in the design of compression members. Upon investigating the test data by DeWolf (1973) it was found (Trestain, 1982) that the effects of local buckling extended into the elastic (Euler) buckling region, neglect of which resulted in unconservative design. This prompted the S136 committee to study the problem in an effort to establish an alternative design approach. The S136 1984 edition uses the tangent-modulus approach, but the interaction effect of local buckling is included over the entire column strength curve.

Combined Axial Load and Bending. A completely new section has been added for singly symmetric sections subject to a combination of axial load and bending about two axes. The method is based on interaction expressions taken from the Rack Manufacturers Institute manual (RMI, 1980).

Wall Studs. The provisions for the design of wall studs were taken directly from the 1980 AISI specification and expressed in the appropriate LRFD format.

Welded Connections

Arc Spot Welds. Explicit design equations for arc spot welds have been incorporated for the first time. Based on research done on behalf of the S136

Technical Committee, design expressions were developed for the shear resistance and tensile resistance of arc spot welds. These equations are dependent only on the sheet steel thickness.

Fusion Welds. The requirements for fusion-welded connections contain provisions formulated from research undertaken in the United States. This work was sponsored by AISI and performed by Pekoz and McGuire (1979). It includes provisions covering fillet welds and flare bevel groove welds.

Connections Made by Bolts, Screws, or Solid Rivets. A number of changes have been made in the design of connections, especially bolted connections. Specific provisions have been provided for the shear resistance of fasteners, the bearing resistance of single fasteners, and for groups of fasteners.

4 Current Specifications

CAN3-S136-M89 (1989a) is a design standard available today which has incorporated the latest research into the behavior of cold-formed structural members. In addition to being state of the art, this standard provides the designer with a specification written in limit state design (LSD) or load and resistance factor design (LRFD) format. In the United States and Europe LRFD specifications are also in preparation.

2.2 STEEL JOISTS

Steel joists are open-web types of beams used to support floors and roofs of buildings. The open-web steel joists first produced in the United States in about 1923 consisted of angles for top chords, rounds for bottom chords, and rounds bent into zigzag patterns for webs to give Warren-type truss profiles. After 5 years without standardization, joist makers organized the Steel Joist Institute (SJI) to standardize and promote uses of steel joists. SJI's first standard specifications for steel joists were published in 1928 and were followed by load tables in 1929.

Table 2.3 Span, depth, weight, and load ranges of SJI open-web steel joists and joist girders *(SJI 1988)*

Type	Span, ft	Depth, in.	Weight, lb/ft	Load, lb/ft	
				Total	Live*
Short span, K-series	8–60	8–30	5.0–17.6	550–127	480–64
Long span, LH-series	25–96	18–48	10–47	927–174	624–83
Deep long span, DLH-series	89–144	52–72	25–70	706–208	395–121
				Panel point load, 1000 lb	Number of joist spaces
Joist girders	20–60	29–72	13–98	4–20	3–12

* Live load limited by deflections not to exceed 1/360 of span length.

SJI's latest standard specifications with load and weight tables for K-, LH-, and DLH-series steel joists and joist girders were published in 1988 (SJI, 1988). Table 2.3 lists span, depth, weight, and load ranges now used. Joist manufacturers are free to choose components. However, SJI often inspects K-series steel makers' facilities before reviewing and approving their outputs. A consulting engineer is retained by SJI to check steel joist designs. British Standards also provide specifications for lattice joists (BS 5950: Part 5, 1987).

1 Chords and Webs

Steel joist designs have been of considerable variety; some examples are presented in Fig. 2.3. Deep long-span steel joists (DLH series) are usually made of hot-rolled angles for both chords and web. Long-span steel joists (LH series) of medium depths are likely to have angles for top chords, rounds for bottom chords, and zigzag rounds for webs. Short-span steel joists (K series) with cold-formed steel top and bottom chords, or rounds for bottom chords and zigzag rounds for webs, have been common since about 1960.

2 Steels

SJI recommends using steels specified by the American Society for Testing and Materials (ASTM), as listed in Table 2.4. ASTM designation A36 (248 N/mm^2) steel is probably the main structural grade used in buildings and bridges in the United States. However, other ASTM designations of high-strength low-alloy steels having specified minimum yield strengths of 248, 310, and 344 N/mm^2 (42, 45, and 50 ksi) are also used in steel joists.

3 Design Guidelines

The designer strives to create steel joists to safely carry the required span loads without excessive weights or costs. Certain structural requirements must be met in determining a joist's rated span load capacity as follows:

1. Allowable design stresses shall not be exceeded.
2. Deflections shall not exceed $^1/_{360}$ of span length in floors and in roofs to which plastered ceilings are attached. Deflections not exceeding $^1/_{240}$ of span length shall be permitted in other roofs.
3. Webs shall safely resist shear loads along spans.
4. Lateral bracing shall be provided at proper span intervals.
5. Compression chords shall be designed for column lengths between panel/point centers. Top chord interior panels of DLH- and LH-series steel joists may be designed at 75% of such column lengths.
6. Compression chords of K-series steel joists longer than 610 mm (24 in.) shall be designed for combined column and bending stresses. All compression chords of LH- and DLH-series joists shall so be designed.

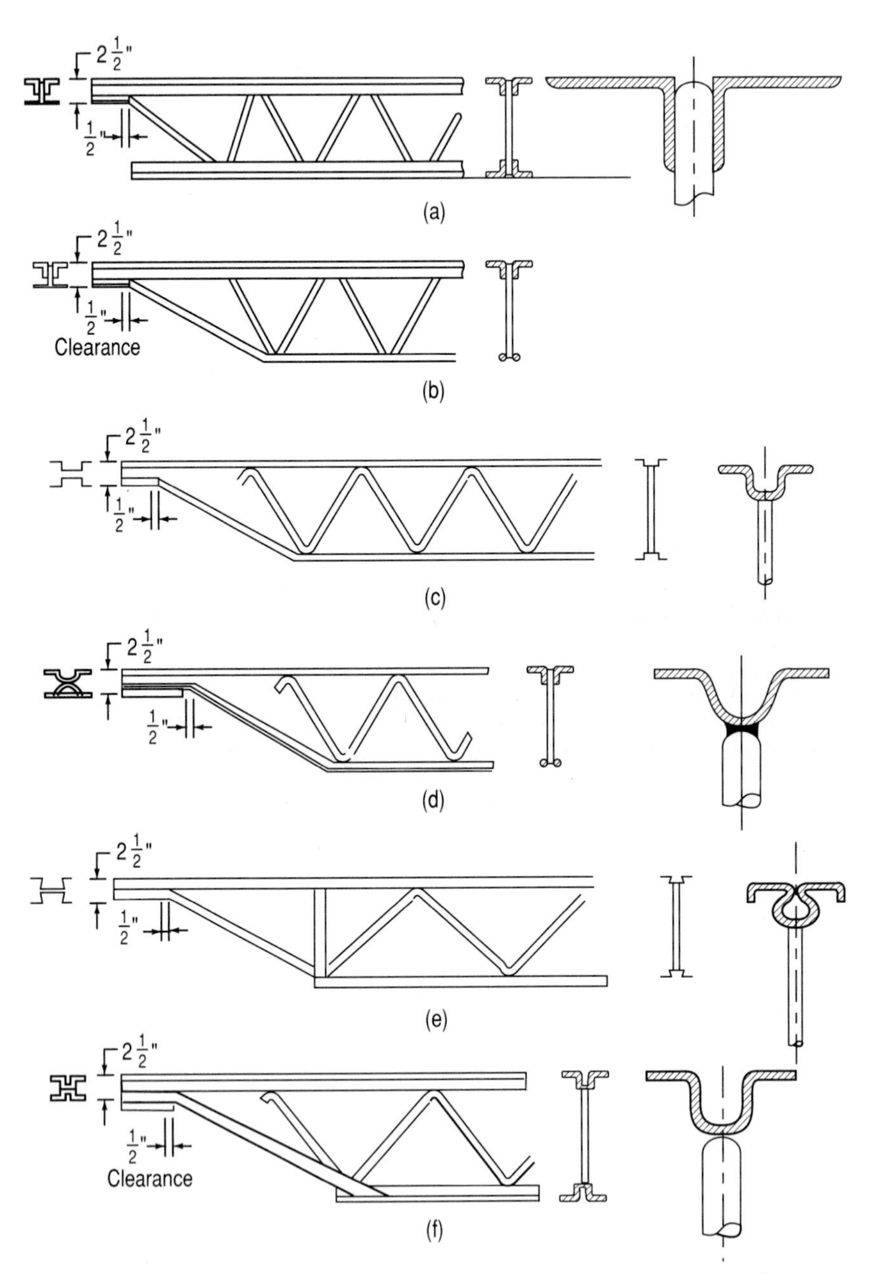

Fig. 2.3 Open-web steel joist profiles.

4 Allowable Design Stresses

SJI's recommended allowable design stresses conform to those specified by AISC (1989). Table 2.5 lists design stresses for K-, LH-, and DLH-series steel joists and joist girders.

Specific values given for K-series steel joists are expressed as factors of specified minimum yield strengths of 248, 290, 310, and 345 N/mm^2 (36, 42, 45, and 50 ksi).

5 Maximum Slenderness Ratios

Maximum slenderness ratios set by SJI are listed in Table 2.6. The SJI column formulas used for the determinations are as follows:

$$C_c = \sqrt{\frac{2\Pi^2 E}{QF_y}} \tag{2.4}$$

For $KL/r > C_c$,

$$F_a = \frac{12\Pi^2 E}{23(KL/r)^2} \tag{2.5}$$

for $KL/r < C_c$,

$$F_a = \frac{[1 - \frac{1}{2}[(KL/r)/C_c]^2]QF_y}{\frac{5}{3} + \frac{3}{8}[(KL/r)/C_c] - \frac{1}{8}[(KL/r)/C_c]^3} \tag{2.6}$$

in which K is the column effective length ratio and Q is the section effective area ratio.

Assuming that all elements of chord sections are fully effective, an L/r value of 90 for an interior chord segment results in an allowable compression stress of

Table 2.4 ASTM steel designations specified by SJI for open-web steel joists and joist girders *(SJI, 1988)*

Short span, K series	Long span, LH series	Deep long span, DLH series	Joist girders
A36	A36	A36	A36
A242	A242	A242	A242
A441	A441	A441	A441
A570	A570	A570	A570
A572 Gr. 50	A572 Gr. 42, 45, 50	A572 Gr. 42, 45, 50	A572 Gr. 42, 45, 50
A588	A588	A588	A588
A606	A606	A606	A606
A607 Gr. 50	A607 Gr. 45, 50	A607 Gr. 45, 50	A607 Gr. 45, 50
A611 Gr. D	A611 Gr. D	A611 Gr. D	A611 Gr. D
Specified minimum yield strength range available, ksi			
36–50	36–50	36–50	36–50

117 N/mm² (16.94 ksi), an L/r value of 120 for an end chord segment gives 72 N/mm² (10.37 ksi), and an L/r value of 200 for web members acting in compression gives 26 N/mm² (3.73 ksi). The AISC upper limit of $L/r \approx 200$ for compression members is thus upheld and the lower L/r values of 90 and 120 set by SJI for chords are desirably conservative. The L/r value of 240 set by SJI for chords and webs acting in tension also confirms AISC's upper limit.

Table 2.5 Allowable design stresses (ksi) specified by SJI for open-web steel joists and joist girders

Type	Specified minimum yield strength F_y, ksi	Tension		Bending			Bearing plates	Compression
		Chords	Webs	Chords	Round webs	Others webs		
Short span, K series	36		22		32.5	22	27	F_a
	50	30	30	30	45	30	37.5	F_a
				Yield strength factors*				
Long span, LH series	0.6	0.6	0.6	0.6	0.9	0.6	0.75	F_a
Deep long span, DLH series	0.6	0.6	0.6	0.6	0.9	0.6	0.75	F_a
Joist girders	0.6	0.6	0.6	0.6	0.9	0.6	0.75	F_a

* Yield strength factors apply to specified minimum yield strengths of 36, 42, 45, and 50 ksi.

Table 2.6 Maximum slenderness ratios L/r specified by SJI for designing chords and webs of open-web steel joists and joist girders *(SJI, 1988).*

Type	Compression			Tension, chords and webs
	Chord segment, interior	Chord end	Webs	
Short span, K series	90	120	200	240
				—
Long span, LH series	90*	120	200	240
Deep long span, DLH series	90*	120	200	240
Joist girders	90	120	200	240
	Maximum allowable compression stress F_a, ksi			Maximum allowable tension stress F_t, ksi
	16.94†	10.37†	3.73‡	30

* An effective length factor of $K = 0.75$ may be used when L is the distance between panel point centers and r is taken as the least radius of gyration for LH- and DLH-series joists.

† Using SJI short column formula when $F_y = 50$ ksi, $Q = 1$, $K = 1$, $C_c = 107$.

‡ Using SJI long column formula when $K = 1$, $E = 29{,}000$ ksi.

6 Design Formulas

Open-web steel joists are designed for bending strength, like other trusses. The following formulas can be used to determine the resisting moment in bending of an open-web steel joist:

$$M = A_c F_c d \tag{2.7}$$

$$M = A_t F_t d \tag{2.8}$$

where M = resisting moment
A_c = compression chord cross-sectional area
A_t = tension chord cross-sectional area
F_c = allowable compression stress
F_t = allowable tension stress
d = distance between compression and tension chord centroids

Equations 2.7 and 2.8 apply regardless of whether the effective compression and tension cross-sectional areas are equal or unequal if Eq. 2.9 is also true,

$$F_c A_c = F_t A_t \tag{2.9}$$

The following formula for the moment of inertia of an open-web joist can be used for combinations of compression and tension chord cross-sectional areas:

$$I = \frac{A_c A_t d^2}{1.15(A_c + A_t)} \tag{2.10}$$

in which I is the moment of inertia.

The factor 1.15 reduces the moment of inertia by an amount which tests on open-web steel joists indicate as being necessary to compensate for the influence of strain in open webs.

7 Bridging

Either horizontal or diagonal bracing may be used for lateral support of open-web steel joists. Spacing of bridging along the span depends on the compression chord size, its stress, and the span length. For K-series steel joists one row of bridging suffices for a 7.3-m (24-ft) span, but five rows would be needed for an 18.3-m (60-ft) span.

8 SJI Load Tables

SJI (1988) standard load tables provide total safe uniformly distributed load-carrying capacities for open-web steel joists in pounds per linear foot of span. These tables are set up in terms of chord sizes, spans, depths, and weights for K-, LH-, and DLH-series steel joists as well as for joist girders. The spans range from 8 to 144 ft (2.44 to 43.9 m), depths from 8 to 72 in. (203 to 1830 mm), and weights from 5 to 70 lb/ft (7.44 to 104 kg/m). The two sets of load values listed in the steel joist tables give total loads [927 to 127 lb/ft (1380 and 189 kg/m)] as well as live loads, when deflection is limited to 1/360 of span length [624 to 64 lb/ft (930 to 95 kg/m)].

9 Cold-Formed Chords

Short-span steel joists with cold-formed chords were introduced around 1960, after it was realized that the low yield strength of annealed mild steel was increased to over 350 N/mm^2 (50 ksi) as the result of the cold work of roll forming. This economical way of achieving high yield strength eliminated the need for buying high-strength low-alloy steels at extra cost. The immediate result was to raise the loading values which could be claimed for mild steel joists by following the procedure outlined in the AISI specification for the design of cold-formed steel structural members (AISI, 1986).

A number of steel producers soon built elaborate machines to roll-form steel chords for resistance welding to solid-bar steel webs on the assumption that such a mass-produced product would be cheaper and more competitive. However, certain factors mitigated against the economics of mass-producing steel joists in captive shops of steel producers:

1. Steel joists orders were highly varied as to span and loading and in end and other details. This made production labor-intensive due to the necessity for frequent machine adjustments. Moreover, labor costs in captive shops of steel producers were higher than those in independent shops.
2. Competitors began making open-web steel joists with hot-rolled steel angles for chords and hot-rolled solid round steel bars for webs in simple jigs, using manual arc welding for joining. They set up their shops where labor costs were low.

10 Applications

Figure 2.4 shows how centering can be fastened to open-web joists preparatory to pouring a concrete floor slab. Figure 2.5 illustrates how open-web steel joists can be anchored to a masonry wall. Open-web steel joists are frequently used to

Fig. 2.4 Centering being fastened to open-web steel joists.

support roofs spanning concrete block walls. Reinforced concrete is the main competitor of steel joists for supporting floors in single- and multistory buildings. However, open-web steel joists do have advantages of lightness and cost against hot-rolled steel shapes such as I-beams in such applications.

11 SJI and Other Research

SJI has sponsored considerable research on open-web steel joists starting around 1965 on such subjects as compression chord design, spacing of bridging, ponding loads, fire-resistance assemblies, vibration, uplift loads, and welding. SJI technical digests summarizing these investigations include Galambos (1970a, 1970b), Heinzerling (1971), Cohn (TD4-1972), Galambos (TD5 Revised-1988), Galambos (1978), Sprague (1978), and Somers and Galambos (1983). Independent research on open-web steel joists has been done in Canada (Robinson et al., 1978) and Australia (Schmidt and Roach, 1973).

12 SJI's 1988 Revisions

In the January 1986 issue of *Civil Engineering,* SJI announced the coming publication of a new set of steel joist standards, including the replacement of the

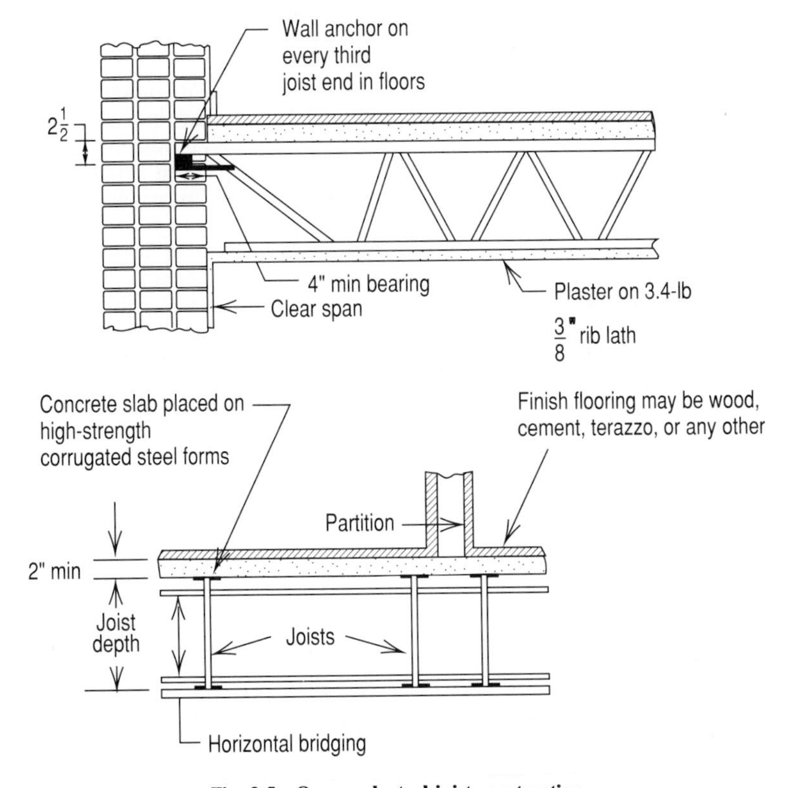

Fig. 2.5 Open-web steel joist construction.

H-series open-web steel joists with a new K-series open-web steel joist. Some features of the *Standard Specifications, Load Tables, and Weight Tables for SJI Steel Joists and Joist Girders* (SJI, 1988) follow.

1. More uniform load capacity differences from one chord section to another
2. Joists which are specifically designed for the lighter loads encountered with standing-seam and membrane roofing systems
3. A total of 64 K-series joists for a wider range of loadings
4. A new economy table to ensure simple and precise selection of the most economical joist
5. New bridging specifications
6. New welding specifications

13 Summary

Open-web steel joists remain important building components in the United States for supporting floors and roofs. Considerable technical information on open-web steel joists is available from SJI, representing the industry, which continues to improve steel joists and to regulate their use. The first open-web steel joists were made of hot-rolled angles for top chords and round bars for bottom chords and web in 1923. SJI organized in 1928 and published its first load table in 1929. The latest standard specifications by SJI are dated 1988.

Cold-formed chords of mild steel utilizing as-formed yield strengths exceeding 344 N/mm^2 (50 ksi) were introduced in 1960. Elaborate machines were built for cold-forming chords to be resistance-welded by using mass-producing techniques. This strategy failed because of the need to supply small, varied orders, requiring frequent machine resettings with consequent high labor costs, especially in shops owned by steel producers.

The current trend in open-web steel joist manufacture is to use hot-rolled angles of yield strengths up to 344 N/mm^2 (50 ksi) for top chords and hot-rolled rounds having yield strengths of 248 N/mm^2 (36 ksi) for bottom chords and webs. Manual arc welding is being used to join steel joist components. Steel joist shops now tend to be established in small towns and rural areas where labor costs have been found to be less. However, some steel joists with cold-formed chords are still being produced.

2.3 CONDENSED REFERENCES / BIBLIOGRAPHY

AISC 1989, *Specification for the Design, Fabrication and Erection of Structural Steel for Buildings*

AISI 1980, *Specification for the Design of Cold-Formed Steel Structural Members*

AISI 1986, *Specification for the Design of Cold-Formed Steel Structural Members*

AISI 1991a, *Load and Resistance Factor Design Specification for Cold-Formed Steel Structural Members*

AISI 1991b, *Commentary on the Load and Resistance Factor Design Specification for Cold-Formed Steel Structural Members*

Allen 1975, *Limit States Design—A Probabilistic Study*

Allen 1981, *Limit States Design: What Do We Really Want?*

BS 5950: Part 5 1987, *Structural Use of Steelwork in Building, Part 5: Code of Practice for Design of Cold Formed Sections*

CAN3-S136-1974 1974, *Cold-Formed Steel Structural Members*

CAN3-S136-M84 1984, *Cold-Formed Steel Structural Members*

CAN3-S136-M89 1989, *Cold-Formed Steel Structural Members*

Cohn TD4 1972, *Design of Fire-Resistive Assemblies with Steel Joists*

CSA-S408-1981 1981, *Guidelines for the Development of Limit States Design*

DeWolf 1973, *Local and Overall Buckling of Cold-Formed Compression Members*

Eurocode 3 Annex A 1991, *Cold-Formed Steel Sheeting and Members*

Fox 1983, *Predicting the Proneness of Buildings to Gross Errors in Design and Construction*

Galambos 1970a, *Design of Compression Chords for Open Web Steel Joists*

Galambos 1970b, *Spacing of Bridging for Open Web Steel Joists*

Galambos 1978, *Structural Design of Steel Joist Roofs to Resist Uplift Loads*

Galambos TD5 1988, *Vibration of Steel Joists—Concrete Slab Floors*

Heinzerling 1971, *Structural Design of Steel Joist Roofs to Resist Ponding Loads*

Hetrakul 1978, *Structural Behaviour of Beam Webs Subjected to Web Crippling and a Combination of Web Crippling and Bending*

Kennedy 1980, *Limit States Design of Steel Structures—Performance Factors*

LaBoube 1978a, *Structural Behaviour of Beam Webs Subjected to Bending Stress*

LaBoube 1978b, *Structural Behaviour of Beam Webs Subjected Primarily to Shear Stress*

NBCC 1985, *National Building Code of Canada, 1985*

Nowak 1979, *Practical Code Calibration Procedures*

Parimi 1976, *Limit States Basis for Cold-Formed Steel Design*

Pekoz 1979, *Welding of Sheet Steel*

Phung 1978, *Structural Behaviour of Transversely Reinforced Beam Webs*

RMI 1980, *Industrial Steel Storage Rack Manual*

Robinson 1978, *Composite Open-Web Joists with Formed Metal Floor*

Schmidt 1973, *Lateral Buckling of Open-Web Joists*

SJI 1988, *Standard Specifications, Load Tables, and Weight Tables for Steel Joists and Joist Girders*

Somers 1983, *Welding of Open-Web Steel Joists*

Sprague 1978, *50-Year Digest—A Compilation of SJI Specifications and Load Tables from 1928 to 1978*

Trestain 1982, *A Review of Cold-Formed Steel Column Design*

Wing 1981, *Web Crippling and the Interaction of Bending and Web Crippling of Unreinforced Multi-Web Cold-Formed Steel Sections*

3

Decking and Sandwich Panels

This chapter reviews uses, good practice, and design recommendations for sandwich panels and decking in high-rise buildings. Sandwich panels are defined as load-bearing elements consisting of two metal skins with a thick core, having low strength and density. Decking is defined as thin-walled sheets of profiled steel especially adapted for use in floors or roofs. Floor decking may be used as shuttering (formwork) or as an integral part of a composite slab. A composite slab is a floor consisting of decking, concrete, and additional reinforcement (Fig. 3.1). Decking is used as a working platform, as shuttering, and finally as permanent positive-moment reinforcement for the slab itself.

3.1 FLOOR AND ROOF DECKING

This section has been structured from the point of view of floor decking, since in tall buildings the most important application for decking is as composite floor slabs. Rules and good practice for composite floor slabs are presented first. This includes decking used as shuttering and as permanent positive-moment reinforcement. Additional practices specific to roof decking are then outlined. The composite action of slabs is not within the scope of this Monograph. For further information on this subject the reader is referred to Council on Tall Buildings, Group SB (1979, chap. 9) and Council on Tall Buildings (1983).

For the United States, good practice, design specifications, and recommendations included in this section are based on the following publications:

Steel Deck Institute (SDI) recommendations (1987a, 1989)

American Society of Civil Engineers (ASCE) specifications (1984)

American Iron and Steel Institute (AISI) specifications (1986)

The opinions expressed in Heagler (1987) represent interpretations and comparisons with SDI recommendations.

For Europe, good practice, design specifications, and recommendations are based on the following publications:

Eurocodes (Eurocode 3 Annex A, 1991, Eurocode 4 Part 1, 1990)

European Convention for Constructional Steelwork (ECCS) documents (1983)

National recommendations such as provided by the Steel Construction Institute (SCI) in the United Kingdom (Lawson et al., 1990) and Samenwerkingsverband Industrie Staalplaat-betonvloeren (SIS) in the Netherlands (1991)

In addition, recent information, which is not being reviewed here, is available in Kozak (1991).

An effort has been made to standardize terminologies and to eliminate overlapping material where possible. Where conflicting opinions are expressed, particular note has been made.

Lastly, it should be noted that a European good practice document for composite slabs is now in preparation at ECCS. The aim of this document is to standardize rules and practices relevant to composite slabs throughout the European region. The SDI will also soon publish a design handbook for composite slab systems; a standard design rationale will be part of this publication.

Decking is often used as a temporary (during construction) or permanent shear diaphragm. Information on this subject is contained in Chapter 4. Information on the acoustic insulation, thermal insulation, and fire resistance of decking is contained in Chapter 6.

3.2 COMPOSITE SLAB APPLICATIONS

1 Materials

Some general information about materials given in this chapter is based on European practice (Eurocode 4 Part 1, 1990; Eurocode 3 Annex A, 1991; prEN

Fig. 3.1 Composite slab.

10 147, 1989; Lawson et al., 1990), which is not essentially different from U.S. practice.

Decking

Steel Grades. The steel grades commonly used in Europe are listed in Table 3.1. There is a tendency to use higher steel grades.

Coating. Decking is commonly fabricated from hot-dipped galvanized plate with a zinc coating of 275 g/m^2 (1 oz/ft^2) on both sides, which corresponds to a mean thickness of approximately 20 μm on each side. This zinc coating is in accordance with Eurocode 4 Part 1 (1990) and is normally sufficient for internal floors in a nonaggressive environment. The steel core thickness varies from 0.70 mm (0.03 in.) to approximately 1.25 mm (0.05 in.).

The use of aluminum-zinc coatings for floor decking should be considered carefully because the chemical reaction between aluminum and cement could have a negative effect on the longitudinal shear capacity of the composite slab after hardening of the concrete.

Concrete

Normal and Lightweight Concrete Grades. The structural concrete used is termed either normal-weight [2350 to 2400 kg/m^3 (147 to 150 lb/ft^3)] or lightweight [1750 to 1900 kg/m^3 (109 to 119 lb/ft^3) dry density]. Wet densities can be 50 to 100 kg/m^3 (3.1 to 6.2 lb/ft^3) higher. In the United Kingdom lightweight concrete consists of expended pulverized fuel ash pellets, sand, and cement. The cement and water contents are higher than in normal-weight concrete. The concrete grade commonly used is C25/30 (cylinder/cube strength). Props should not be removed until the concrete has gained a minimum of 75% of its required (design) strength. This is often achieved after 7 days.

Aggregate. The nominal size of the aggregate depends on the smallest dimension in the structural element in which the concrete is poured. According to Eurocode 4 Part 1 (1990), it shall not exceed the least of (Fig. 3.2):

$0.40h$

$b_0/3$

31.5 mm (1.22 in.) [sieve C31.5]

The detailing provisions for the slab thickness, indicated in Fig. 3.2, are in accordance with Eurocode 4 Part 1 (1990).

Chloride. All manufacturers warn not to use calcium chloride or other salts in the concrete.

Table 3.1 Steel grades commonly used in Europe* *(prEN 10 147, 1989)*

Grade	Yield strength of basic material, MPa (psi)
FeE220G	220 (32,000)
FeE250G	250 (36,000)
FeE280G	280 (41,000)
FeE320G	320 (46,000)
FeE350G	350 (51,000)

2 Decking Geometries

The historical development of roof decks for buildings presented here was originally written for the United States in Council on Tall Buildings, Group SB (1979) but is typical of floor decking developments throughout the world. It should be noted that economical floor deck geometries are normally different from those used in roofs.

One early type of decking uses transverse wires welded to the decking to provide the shear connection (Fig. 3.3). Composite action is limited by the strength and spacing of the welds and transverse wires. Another type of decking provides interaction between concrete and steel by its unique shape (Fig. 3.4).

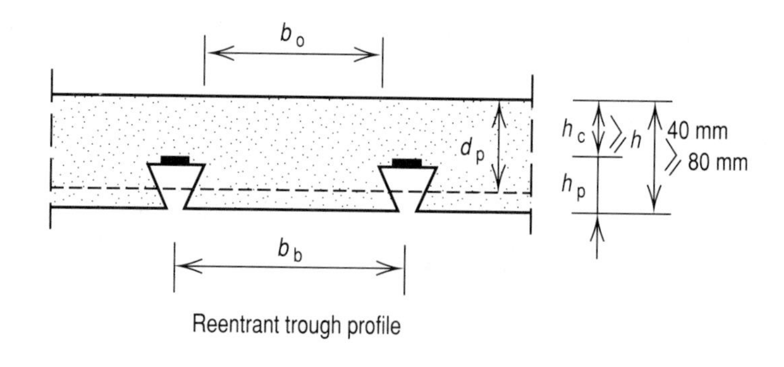

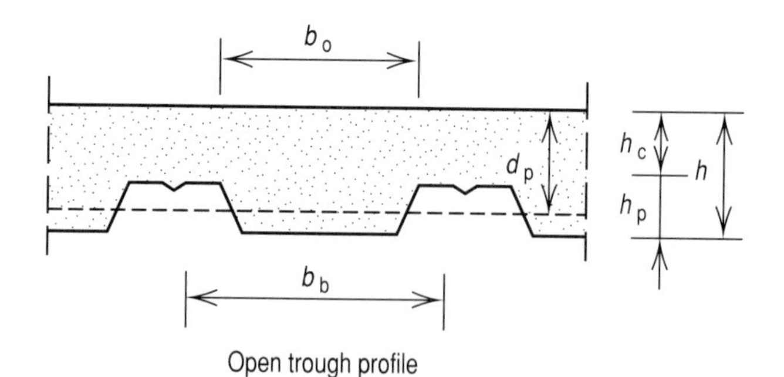

Fig. 3.2 Decking and slab dimensions.

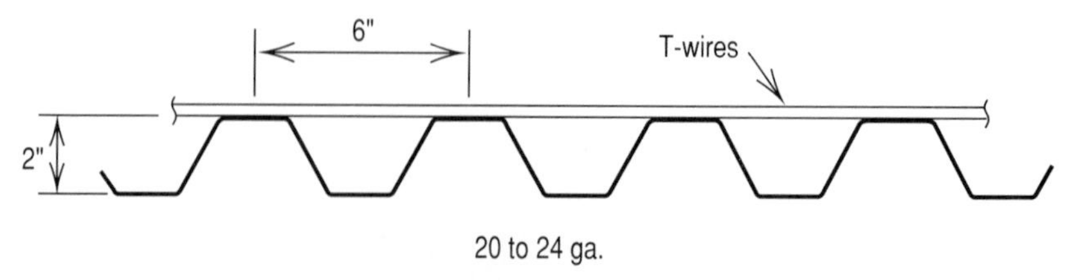

Fig. 3.3 Decking with welded transverse wires.

In the 1960s a third type of decking was introduced which has embossments or deformations cold-rolled into the webs and flanges of ribs, providing mechanical interaction between concrete and steel (Fig. 3.5). This type of decking has become increasingly popular. It is economical to manufacture because there is only a minor additional cost associated with embossing the webs and flanges of decking during the forming operation. Decking configurations are often the same as those used for roofs in noncomposite built-up roofing systems.

Typical decking used in the United States is shown in Fig. 3.6 (Heagler, 1991). The 38-mm (1.5-in.) deck shown is the same as the roof deck wide-rib profile, which is the result of double-duty use of existing tooling. Often the 38-mm (1.5-in.) deck is installed in the inverted position so that more concrete surrounds the shear studs welded to the beams.

Decking can be either noncellular or cellular. Cellular systems consist of deckings with flat steel plates or two identical deckings held together as illustrated in Fig. 3.7. There is no essential difference between noncellular and cellular

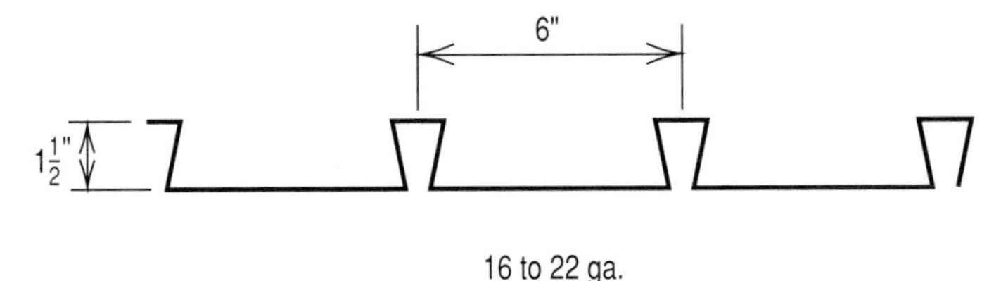

Fig. 3.4 Unique shaped deck.

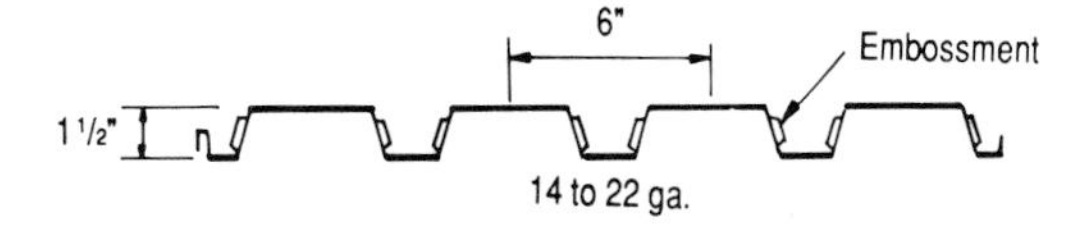

Fig. 3.5 Deck with embossments.

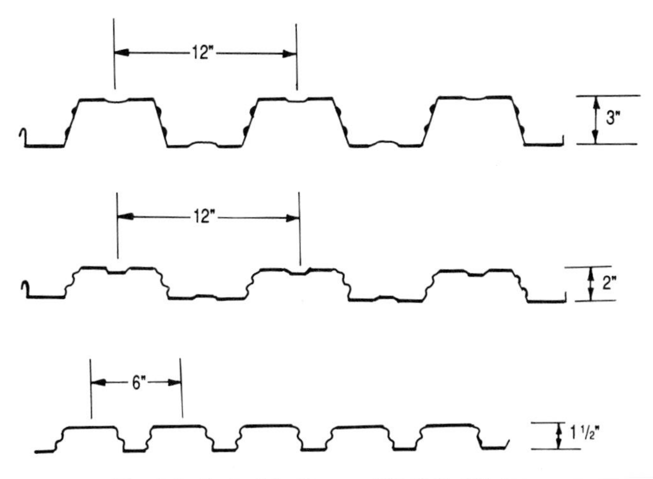

Fig. 3.6 Typical decking used in United States.

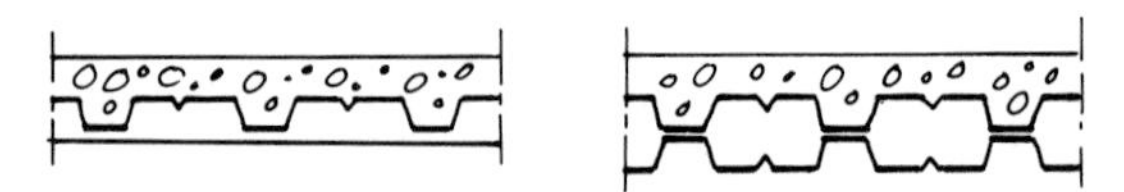

Fig. 3.7 Examples of cellular systems.

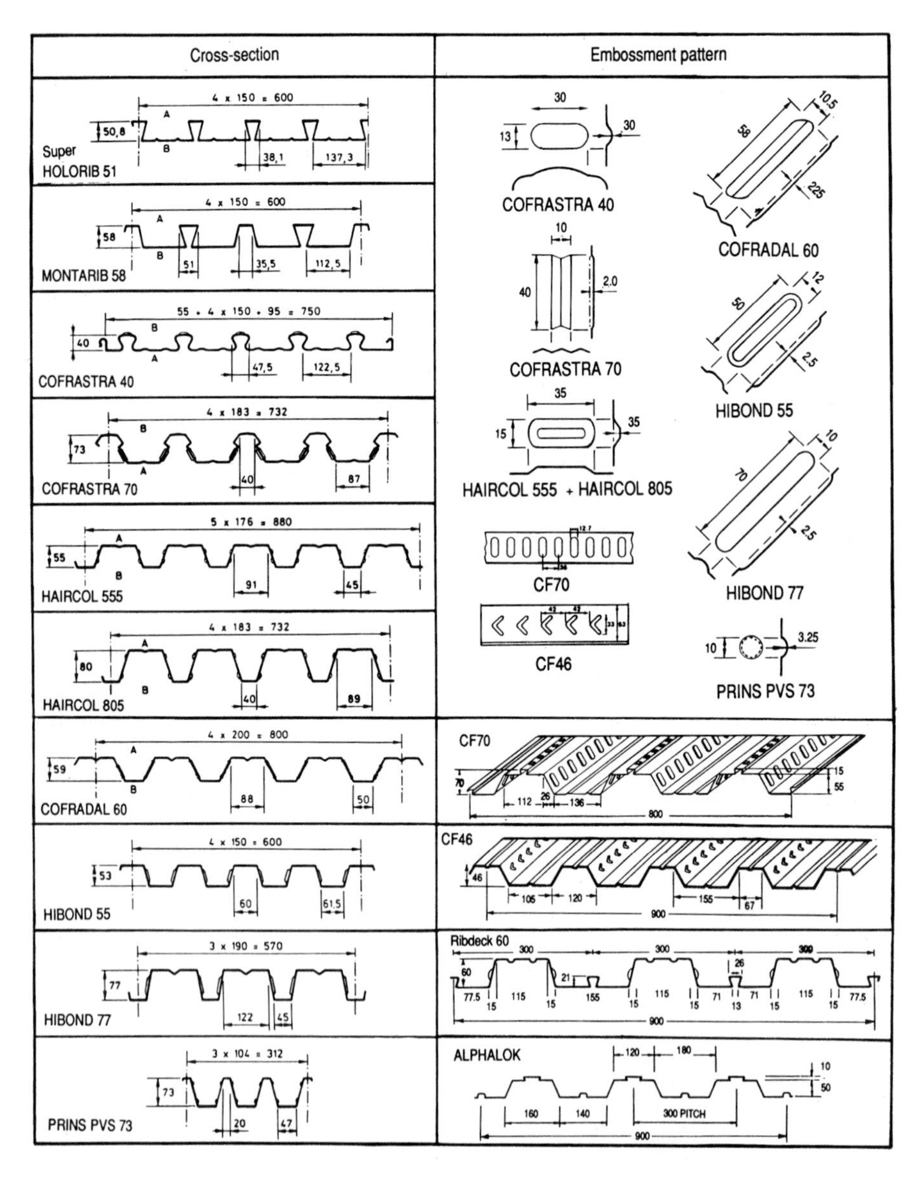

Fig. 3.8 Typical decking used in Europe *(Daniels, 1990).*

systems in their composite action. Cellular systems are seldom used in Europe. Therefore this chapter only deals with noncellular systems. Some typical decking used in Europe is shown in Fig. 3.8 (Daniels, 1990).

3 Functions

Decking must perform three separate functions. First, the deck is used as a working platform; it must support the workers involved in various trades and, in some cases, stored materials. The second function is to act as concrete formwork. Lastly the decking is the positive bending reinforcing for the composite floor. Other important contributions of decking may also be required by the designer, such as diaphragm bracing or acting as part of a fire barrier.

Fig. 3.9 Installation of metal decking (*Photo Courtesy: TNO Building and Construction Research*).

4 Advantages

The possible advantages of floor decks over alternative floor systems are discussed in this section. These advantages remain essentially valid for most countries (Council on Tall Buildings, Group SB, 1979).

Decking (Fig. 3.9) is lightweight and, therefore, easily handled and placed. This reduces installation time and minimizes site labor. The construction time is reduced since casting of additional floors may proceed without having to wait for previously cast floors to gain strength to support the formwork. The decking provides positive-moment reinforcement for the composite floor. Usually only a minimum amount of crack reinforcement is needed.

The use of decking can simplify the problem of scheduling various trade unions on the job, and can reduce jurisdictional disputes between unions. In the United States steel forms are usually placed by ironworkers, who also erect the structural steel work and field-weld shear connectors (Fig. 3.10). Carpenters may be needed only for forming around columns and openings for minor subdivision of slab formwork and possibly for placing shoring. The forms are left in place permanently.

Fig. 3.10 Cold-formed metal deck installation of shear connectors with beams (*Photo Courtesy: TNO Building and Construction Research*).

Decking can easily be combined with preengineered raceways for electrification, communication, and air distribution systems. It can serve as a ceiling surface or provide for easy attachment of a suspended ceiling. An additional advantage is that the decking, as soon as it is fixed in place, can act as an effective in-plane shear diaphragm. Furthermore, in general, composite slabs are lighter than other floor systems. This can result in substantial savings in foundation costs. Finally, for the composite slab 2-hr fire ratings can easily be achieved (see Subsection 5 in Section 3.3).

3.3 DESIGN ASPECTS OF COMPOSITE SLABS

1 Loading

Loads Prior to Concreting. According to Heagler (1987) no loading specifications have been published in the United States for decking subjected to loads *prior* to concreting. This is somewhat surprising; it is an important function. He further reports that some specifications call for the use of planking under stored materials and in walkway areas, but seldom is an actual design load listed. In the United Kingdom and the Netherlands the floor decking is designed such that it performs adequately when subjected to the following loads prior to concreting, as recommended in Lawson et al. (1990) and adopted in SIS (1991). When the decking is fixed in place, the deck often will be used for storage and as a working platform. This is acceptable if the total load on the decking is not more than 3 to 4 kPa (63 to 83 psf).

The largest loads in the construction stage can be considerable. Table 3.2 gives an overview of a number of frequent loads. In the case of very large loads or if there is any doubt about sufficient load-carrying capacity of the decking, the loads should preferably be placed just above a supporting beam (Fig. 3.11). Wooden beams should be used to spread local loads. Movable loads need special attention.

Loads During Concreting (Construction Loads)

Europe. In Europe floor decking must be designed such that it performs adequately when subjected to the construction loads, as specified in Eurocode 4 Part 1 (1990). Loads during concreting are the weight of workers, piled concrete, and pump lines, the self-weight of the concrete, and the like (Fig. 3.12). These loads, taken together, are called construction loads. The construction loads listed in Table 3.3 are recommended in Lawson et al. (1990) and adopted in SIS (1991).

Table 3.2 Loads before pouring concrete *(Lawson et al., 1990)*

Product	Weight
1-m-high stack of mesh reinforcement	2 to 3 kN/m
Standard drum with stud shear connectors	5 kN
Standard 2.5- by 4.5-m skip with lightly packed deck off-cuts, empty drums, or timber	4 to 8 kN/m
Bundle of 100 12-mm-diameter reinforcing bars	1 kN/m
Stack of 20 decks	3 kPa
Wheel load of a standard light compressor	10 kN

The loads given in Table 3.3 can, however, be very local. Loads will be partly carried by adjacent decks, which are not or less loaded. Therefore the construction load is mostly considered to be a uniformly distributed mean value for the entire floor. Eurocode 4 Part 1 (1990) assumes construction load to be equal to a uniformly distributed variable load of 1.5 kPa (31.3 psf) over an area of 3 by 3 m (10 by 10 ft), placed in the most unfavorable position, and a uniformly distributed variable load of 0.75 kPa (16 psf) over the remaining area (Fig. 3.13). These values are not adequate when excessive concrete heaping occurs or when strong impact forces arise, as will be the case when the concrete is poured incorrectly. Deviation from these load values is permitted if it is proven that a lower construction load is acceptable.

In addition to loads that can be considered uniformly distributed over the floor surface, there are also very local loads, such as working on the decking. Decking therefore should be designed for a live load of 3 kN/m (0.21 kips/ft) transverse to the span direction and in the most unfavorable situation, such that the self-weight of the concrete mortar does not need to be taken into account.

The required slab depth should also be considered. Each additional 10 mm (0.4 in.) of concrete means an extra weight of 0.25 kPa (5.2 psf). Excessive deflection of the decking can result in a ponding effect of the concrete, which also means extra weight. As an approximation it can be stated that the weight of the additional concrete thickness is equal to a uniform concrete thickness of $^2/_3\delta$ (Fig. 3.14), δ being the midspan deflection, including the ponding effect caused by the self-weight of the decking and the concrete.

United States. In the United States floor decking must be designed such that

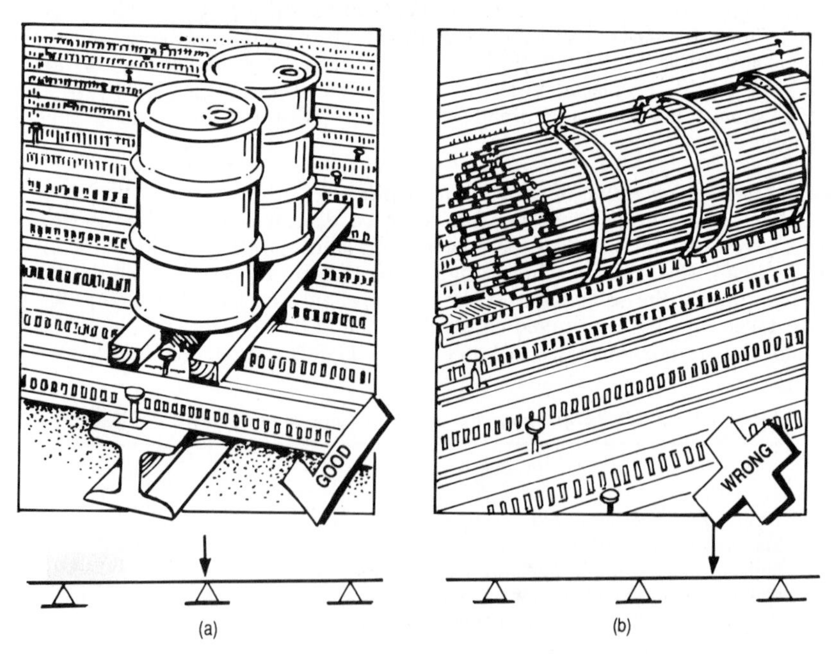

Fig. 3.11 Local loads on decking before pouring concrete. (*a*) Loads above supports. (*b*) Loads in midspan.

it performs adequately when subjected to the following construction loads, as specified by SDI (1987a) and ASCE (1984). These loading criteria are expressed in Heagler (1987).

The subsequent use of a deck as shuttering usually carries a concrete weight of at least 1.44 kPa (30 psf) besides the construction load of 0.96 kPa (20 psf). Thus a

Fig. 3.12 Placing of concrete on cold-formed metal deck (*Photo Courtesy*: *TNO Building and Construction Research*).

load of 2.39 kPa (50 psf) is reasonable for a working platform. It would cause no increase in decking thickness and would serve as a useful limit to the contractor. As a result 2.39 kPa (50 psf) is suggested as the loading for the working platform. During pouring of the concrete, care should be taken to avoid excessive concrete heaping. Ideally concrete should be poured over beams and spread to the center of the deck. For most unshored jobs the deck deflection will cause an increase in concrete volume of less than 5%. The frame deflection will be a much more significant factor, so many designers compensate by calling for specific cambers.

Figure 3.15 shows the loading criteria for a deck used as shuttering (SDI, 1987a; ASCE, 1984). Unshored construction is generally preferred because of costs and speed, so designers generally choose the most economical deck (least metal thickness) that can carry the concrete without intermediate and temporary supports. This policy ensures a load capacity of 2.39 kPa (50 psf) for the working platform. For one span only, SDI has recently applied a factor of 1.5 to W_1. The formulas given in Fig. 3.15 then become

$$+M = 0.125(1.5W_1 + W_2)l^2$$
$$+M = 0.25Pl + 0.188W_1l^2 \tag{3.1}$$

Cantilevering the deck over supports is common practice in the United States. Figure 3.16 shows the suggested loading for checking cantilever spans.

Table 3.3 Some frequent loads during pouring of concrete *(Lawson et al., 1991)*

Load	Weight
Four men in an area of 2 m^2 (21.5 ft^2)	1.5 kPa
Cone of heaped concrete of 0.2-m (0.65-ft) height and 1-m (3.2-ft) base	2 kPa
150-mm (6-in.)-diameter pipeline full of concrete	0.4 kN/m

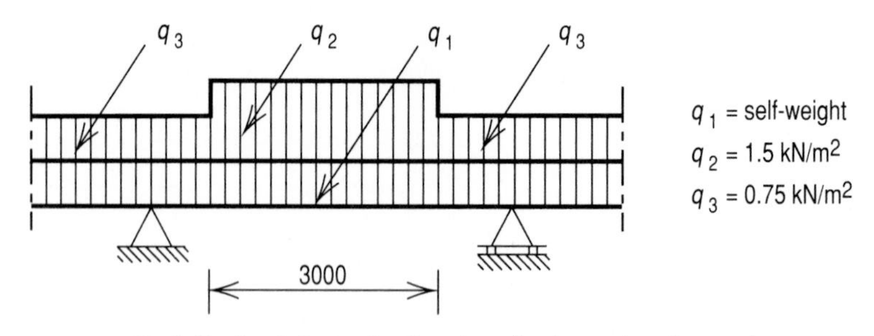

Fig. 3.13 Loads in construction stage due to pouring of concrete.

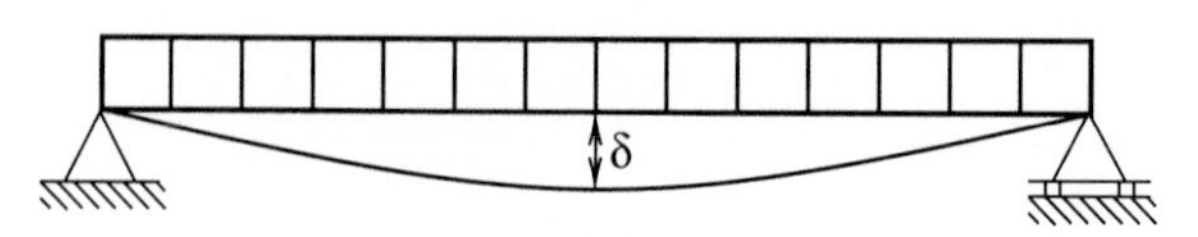

Fig. 3.14 Ponding effect.

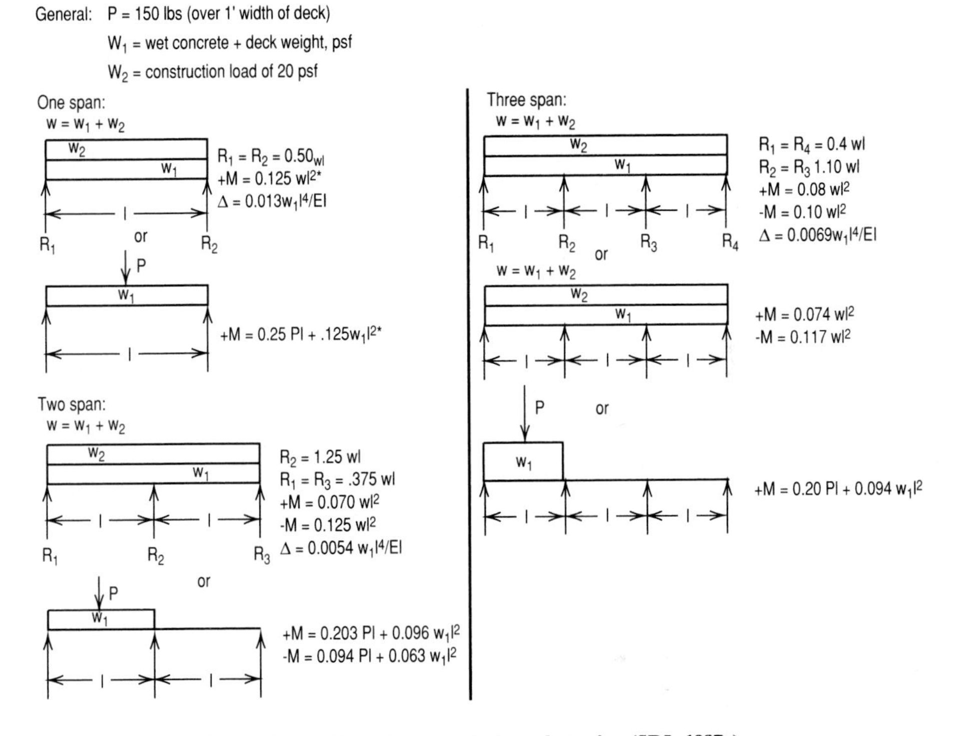

Fig. 3.15 Loading criteria for deck as shuttering (SDI, 1987a).

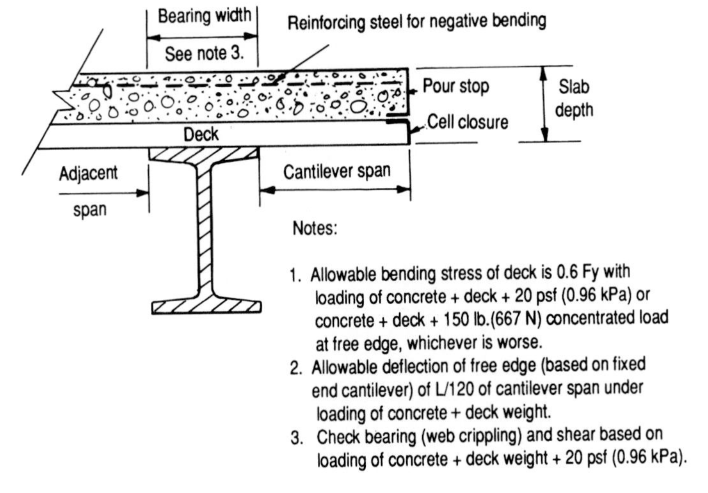

Fig. 3.16 Loading for cantilever spans.

Loads After Concreting. No specifications in the United States are reported by Heagler (1987). In the United Kingdom and the Netherlands the floor decking is designed such that it performs adequately when subjected to the loads after concreting listed in Table 3.4, as recommended in Lawson et al. (1990) and adopted in SIS (1991). After pouring of the concrete, several loads can be placed on the floor. Table 3.4 lists some of the most common loads. Immediately after pouring, the strength of the concrete is negligible. Therefore the load should not exceed the construction load that had been taken into account for the construction stage. Heavy loads can only be placed if the concrete has reached sufficient strength. The attained strength can be determined by concrete cube tests. Heavy loads such as those given in Table 3.4 should also be placed above the supporting beams.

2 Openings

The design of openings in decking is an important consideration that is too often overlooked. Openings come in many sizes and must be defined to include both cutouts and damaged regions of the decking. This section includes the design of decking and possible additional supports, whether or not the cutout is made before or after composite action is achieved.

In the United States Heagler (1987) reports that holes are frequently cut in the deck for various purposes. The size and location of the openings are usually not known until the deck has been installed. Some general guidelines are required, but the fact that size and location are not known beforehand can make an individual study for each opening necessary. The designer can best handle the problem by requiring all openings above a certain size to be reinforced.

Table 3.4 Most common loads after pouring of concrete *(Lawson et al., 1990)*

Load 1 m (3.2 ft) high	Weight
Stack of bags of fire protection material	2.5 kPa
Stack of bags of cement mortar	10 kPa
Pallet bricks	15 kPa

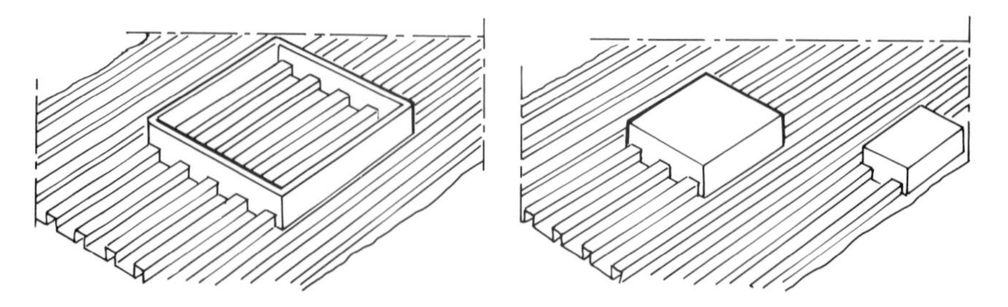

Fig. 3.17 Typical formwork for boxing out openings.

Three typical opening dimensions are discussed. Dutch recommendations for openings (SIS, 1991) have been used as a basis for this discussion. The point of view in the U.S., as expressed by Heagler (1987), is in agreement with the Dutch recommendations.

Openings up to 150 mm (6 in.). These openings, which are often used for conduits, can be made without framing. Reinforcement to spread the loads can also be omitted. Nevertheless it is advisable to consider reinforcement transverse to the direction of the ribs of the decking in order to spread the loads in the composite stage to the adjacent ribs. Round openings of this diameter can be drilled with a diamond bit after hardening of the concrete. The bond between decking and concrete should be guaranteed. In certain cases it is a problem if the decking can be easily lifted off the concrete because of its form; then a reentrant profile behaves very well. If the profile can lift off the concrete easily, the hole in the concrete should be boxed out and the decking snipped (or cut) after hardening of the concrete (Fig. 3.17). In general, boxed out openings of this size should be discouraged because it is very difficult to cut directly along the edges of the openings. (Cutting will damage the tools.) Special elements for finishing cannot be used without additional provisions. On the other hand, openings in the decking which are installed before pouring the concrete can very well be finished with special elements, although from a structural point of view it is not always possible to choose this solution without temporary props. Therefore alternatives are given in the following.

Openings up to 300 mm (12 in.). In most cases these openings can still be made without framing. Reinforcement transverse to the ribs is necessary to spread the load in the composite stage to the adjacent ribs. Depending on location and load capacity, additional longitudinal reinforcement is required in the troughs along the opening (Fig. 3.18*a*).

Some contractors prefer simply to block out the slab (Fig. 3.17) for smaller openings and to cut the deck after the concrete has cured. This is generally a good practice as the working platform is not weakened. Severely dented areas should be treated in the same manner as openings. Openings between 150 and 300 mm (6 and 12 in.) might also be reinforced with a flat plate (18 gage or 1.2 mm minimum) spanning onto adjacent ribs and being welded (or screwed) into place (Fig. 3.18*b, c*).

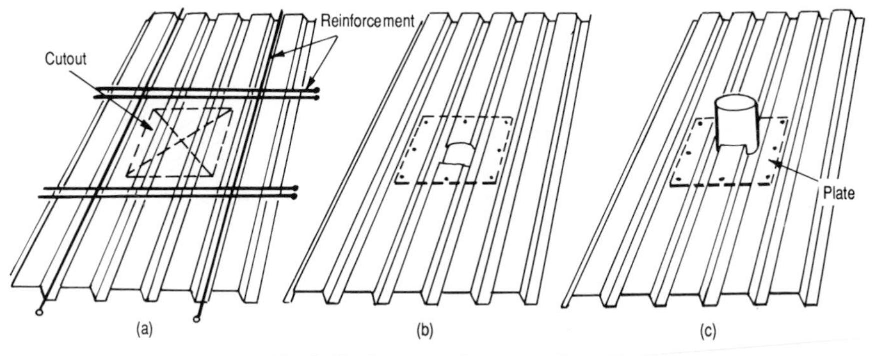

Fig. 3.18 Small openings (one rib).

Openings up to 450 mm (18 in.). For these openings framing is generally necessary, but it is often possible to place the framing in the depth of the floor. Openings of 300 to 450 mm (12 to 18 in.) might be reinforced with structural angles on top of the deck and on each side of the opening, perpendicular to the deck ribs, and spanning onto two adjacent deck ribs on each side (Fig. 3.19*b*). The aim, of course, is to spread the load into adjacent uncut ribs. The "hidden beam" can also be executed with reinforcement depending on:

The position of the openings in the span of the composite floor
The concrete depth on top of the decking
The ratio of the maximum load-carrying capacity to the actual load

In the examples shown in Fig. 3.19, openings are made in the decking after the concrete has hardened. In that case, steel sections transverse to the ribs are put in place to spread the loads in the composite stage to the ribs or the hidden beam in the longitudinal direction along the opening. If the opening is cut before pouring the concrete, the edge of the decking must be supported (Fig. 3.19*c*). In such a case it is often necessary to use a temporary prop in the construction stage.

Openings Larger Than 450 mm (18 in.). Larger openings may require framing on the deck underside to tie into the structural frame and should be analyzed by the engineer (Fig. 3.20). In this case the openings in the decking can easily be made before pouring the concrete.

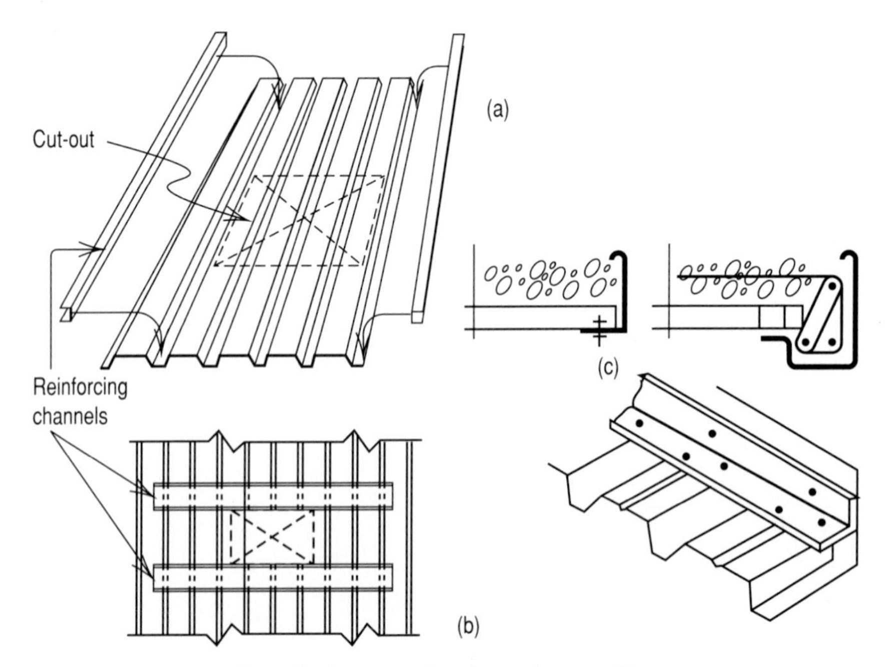

Fig. 3.19 Large openings (more than one rib).

3 Supports

Proper support detailing is essential to the safe use of decking as a working platform or as shuttering. Eurocode 4 Part 1 (1990) specifications are reproduced here. The minimum bearing length for decking on end supports according to Eurocode 4 Part 1 (1990) is

On support of steel or concrete	50 mm (2.0 in.)
On other materials	75 mm (3.0 in.)

The minimum bearing length for continuous decking on an intermediate support is 75 mm (3.0 in.) for steel or concrete and 100 mm (4.0 in.) for other materials (Fig. 3.21). Because of embossments in the web and flange an overlap is not always possible with every decking.

Common practice in the United States is to either butt or lap decking ends. It is usually more practical to butt. Shear studs are best applied through one thickness of decking (except in case of cellular deck), and composite indentations keep the deck from tightly nesting, so firm attachment to the structure is harder to obtain when the deck is lapped. On wide-flange supports covering of the gap may not be necessary. Most gaps that require covering are usually wider than 12 mm (0.5 in.) (Heagler, 1987).

4 Structural Design

The design of decking for use as shuttering, also called formwork, is discussed. When used only as shuttering, the deck must safely carry wet concrete and construction loads. The final service load will be obtained by reinforcing the slab; thus the deck choice influences only the geometry of the slab. Construction load and design for such decking are essentially the same as for decking that will act as part of a composite slab.

Guidelines for Economic Design. Composite floors are most economical if their use is considered in a very early stage—when making the construction plans. The advantages of using decking are affected considerably by the choice of the

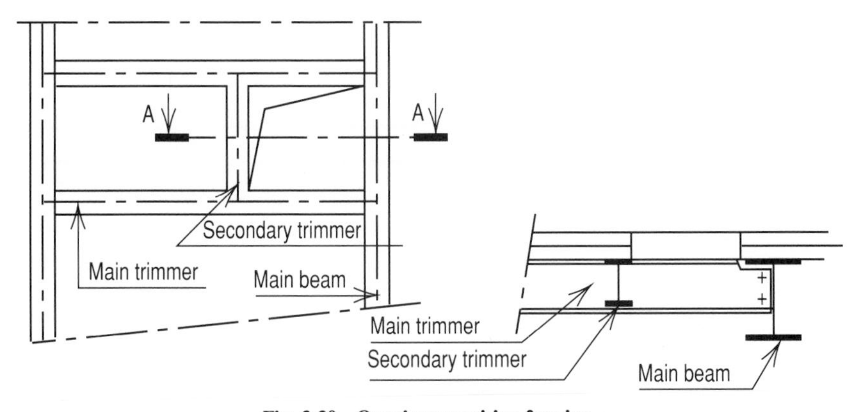

Fig. 3.20 Openings requiring framing.

structural frame. Indeed, the greatest commercial success of floor decking has resulted from the elimination of shoring. If shoring is required, it is necessary to check whether it is required before the deck is installed. In that case worker loads must be considered. Reentrant decking is often favored because of its improved fire resistance. Fire tests have proved that this decking, without additional reinforcement, has better thermal behavior than open trapezoidal decking.

There is some tendency in Europe to use deeper decking. But decking deeper than 80 mm (3.2 in.) is not used frequently. For a structural frame in a grid of 3.6 m (11.8 ft) cost estimates may indicate the following three options (Fig. 3.22):

1. *Present decking in a structure with secondary beams on 3.6-m (11.8-ft) centers* (Fig. 3.22*a*). For this option floor heights must be minimized. An important reduction in materials can be achieved by making the slab composite with all underlying beams.
2. *Deep long-span decking in a structure without secondary beams and a decking placed on the upper flange of the main beam* (Fig. 3.22*b*). Composite action is less optimal because the strength of the shear connectors is reduced (for the same rib width) due to the rib height. Also, the floor height will be greater than in option 1 if subbeams and main beams are placed at the same level. On the other hand, material savings are possible (no secondary beam) and fabrication (less connection details) and erection (less connections) times are reduced considerably.

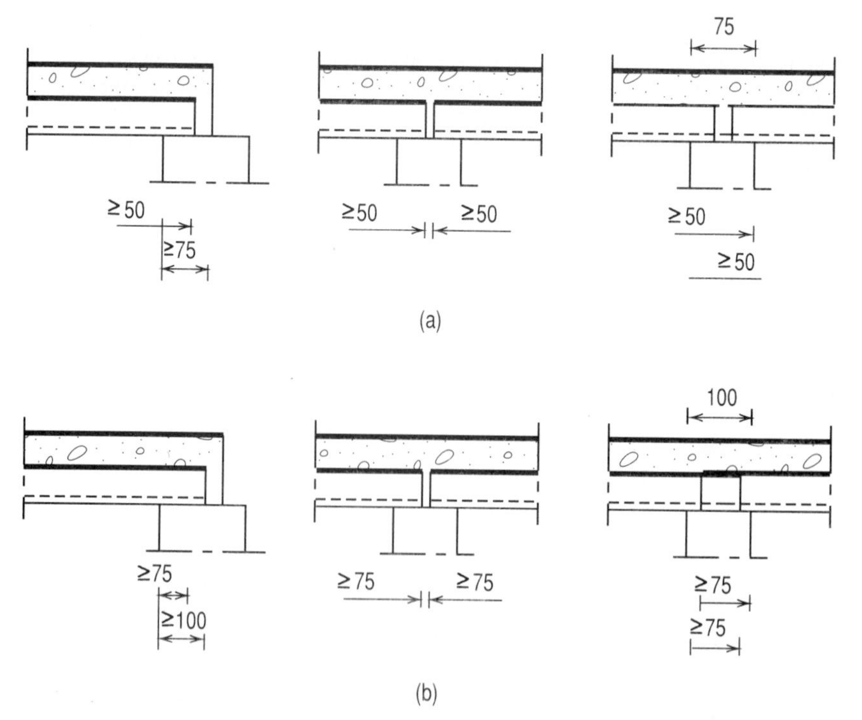

Fig. 3.21 Bearing length. (*a*) Support of steel and concrete. (*b*) Support of other material.

3. *Deep long-span decking in a structure without secondary beams and decking placed on the bottom flange of the main beam* (Fig. 3.22*c*). Composite action between beam and floor slab is limited by the decking rib widths. However, the floor height will be smaller than in option 2. This could result in significant cost savings.

Design Rules and Aids (Tables and Graphs)

Europe. Structural capabilities are determined in Europe by Eurocode 3 Annex A (1991). To determine the capacity of the decking it is possible to use a calculation model or to run tests. The calculation model is based on a schematization of the profile into effective parts of the flange and web. The embossments in the decking could affect the structural behavior. For example, if there are vertical embossments in the web of a decking, an axial force in this part of the web can only be transferred after large deformation. Figure 3.23 shows the remaining effective parts of a decking due to the buckling of compressive zones and the influence of the embossments in the web.

As an example of a design aid in the Netherlands, Table 3.5 gives the maximum decking spans for composite slabs without propping and for a specific slab depth (SIS, 1991). In this table a construction load of 1.5 kPa (31 psf) has been taken into account. Strength and stiffness are in accordance with Eurocode 3 Annex A (1991). In case lightweight concrete is used, longer span lengths can be

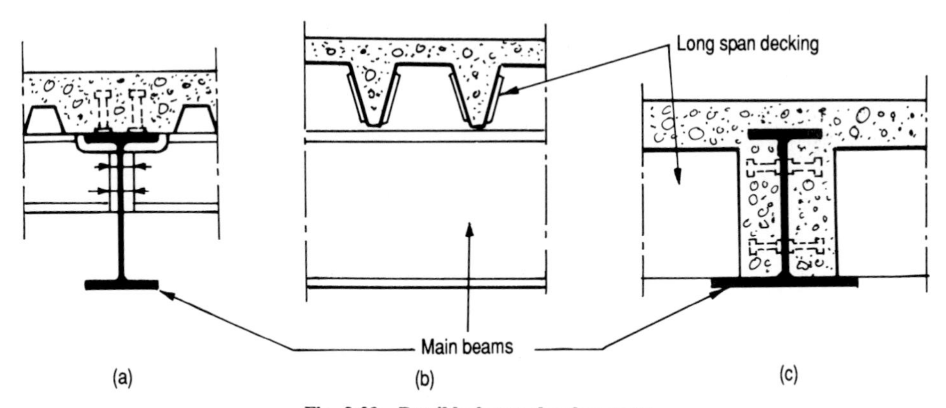

Fig. 3.22 Possible future developments.

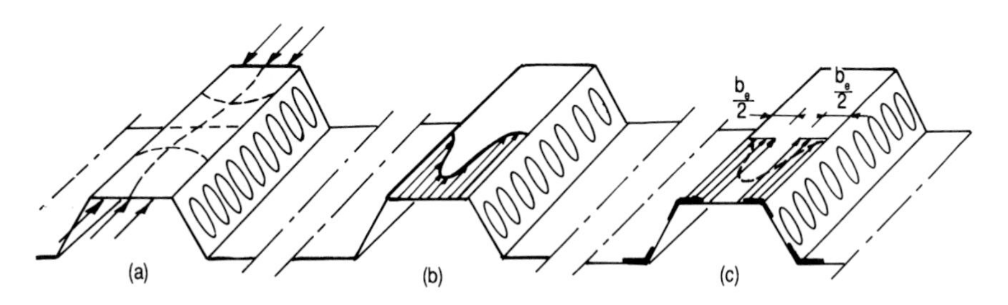

Fig. 3.23 Effective parts of decking.

attained without propping. Therefore it is acceptable that lightweight concrete costs increase slightly.

United States. SDI has issued recommendations for construction loading in the United States as shown in Fig. 3.15. The structural capabilities (strength and stiffness) are determined by AISI (1986). Cold-formed means that the decking is formed into shape at room temperature. The AISI document also provides the acceptable base-steel specifications.

Manufacturers have generally translated the SDI-recommended construction loading into maximum recommended spans or into required section properties. Applicable tables are provided in individual company brochures. Contractors need to know the uniform load capacity of the installed deck. When SDI recommendations for decking design as shuttering are followed, a minimum of 2.39 kPa (50 psf) is available.

The decking may be used in the United States as a diaphragm both with and without the final concrete (SDI, 1987b). The SDI published tables for 63.5- by 12.7-mm (2.5- by 0.5-in.) standard decking covering both conditions. For other decking profiles tables should be prepared by individual manufacturers. SDI also published tables for lightweight insulating fill systems carried by the standard decking.

Decking design in the United States can thus be summarized as follows:

1. Select proper deck to carry the needed slab thickness.
2. Check "bare" deck diaphragm strength and stiffness if there are in-plane construction load requirements. (See Chapter 4 concerning shear diaphragms.)

Table 3.5 Maximum span lengths during construction stage without temporary propping

Decking	Slab depth, mm	Self-weight, kPa	Concrete grade	Maximum span length: Single span	Two spans	Three spans	Four spans
Holorib 38/0.75	90	2.08	17.5	2.10–2.20	2.00–2.10	2.20–2.30	2.10–2.20
Holorib 38/0.88	90	2.10	17.5	2.30–2.40	2.30–2.40	2.50–2.60	2.40–2.50
Super Holorib/0.9	130	3.30	17.5	2.30–2.55	2.55–2.80	2.80–3.05	2.55–2.80
Cofrastra 40/0.75	130	2.96	25.0	—	—	2.00–2.15	2.00–2.15
Peva 45/0.8	95	1.94	17.5	2.30–2.45	2.30–2.45	2.60–2.75	2.45–2.60
CF 46/0.9-grindb.	100	2.03	17.5	2.40–2.50	2.30–2.40	2.50–2.60	2.40–2.50
CF 46/1.2-grindb.	110	2.30	17.5	2.70–2.80	2.90–3.00	3.10–3.20	3.00–3.10
CF 46/0.9-lichtb.	100	1.70	22.5	2.40–2.50	2.40–2.50	2.60–2.70	2.50–2.60
CF 46/1.2-lichtb.	120	2.13	22.5	3.00–3.10	2.90–3.00	3.20–3.30	3.00–3.10
Hi Bond 55/0.75	105	1.98	17.5	2.30–2.45	2.30–2.45	2.60–2.75	2.45–2.60
Hi Bond 55/0.88	115	2.23	17.5	2.60–2.75	2.60–2.75	2.90–3.05	2.75–2.90
QL 59/1.0-grindb.	130	2.63	17.5	3.00–3.10	2.50–2.60	2.80–2.90	2.70–2.80
QL 59/1.0-lichtb.	120	2.00	22.5	2.70–2.80	2.80–2.90	3.10–3.20	3.00–3.10
Cofradal 60/0.75	140	2.59	37.5	2.60–2.70	2.00–2.10	2.20–2.30	2.10–2.20
Cofrastra 70/0.75	125	2.35	25.0	2.60–2.75	2.45–2.60	2.60–2.75	2.60–2.75
PSV73	125	2.48	17.5	2.75–2.90	2.90–3.05	3.20–3.35	2.90–3.05
Super floor 77/0.75	160	2.88	25.0	3.20–3.35	2.15–2.30	2.30–2.45	2.30–2.45
Super floor 77/1.0	145	2.54	25.0	3.05–3.20	3.20–3.35	3.50–3.65	2.90–3.05

3. Check the deck-slab combination for diaphragm capabilities during service loads.
4. If diaphragm behavior in either construction or service is insufficient, try to upgrade the fastening rather than increasing the deck thickness.

For composite slab design in the United States reference is also made to the Council on Tall Buildings, Group SB (1979). Floor decking design for shuttering only follows conventional procedures for reinforced concrete slabs with one important difference. If the decking is galvanized, it must have the same durability as the structure. If the decking is used without shoring, it will always carry the dead load of the slab; the slab can then be reinforced for only the live load requirements. Permanence and durability of the decking should be considered also for diaphragm design.

5 Fire Resistance

During the construction stage, fire resistance measures are generally not necessary. For normal composite slab construction practices, in Europe and the United States no special fire resistance measures are required to obtain a 30-min fire rating. This implies that the slab is capable of supporting its design load during a "normal" fire for 30 min and that the temperature at the top of the floor stays below a maximum of 180°C (356°F). Penetration of hot gases is not a criterion for composite slabs due to the presence of the decking.

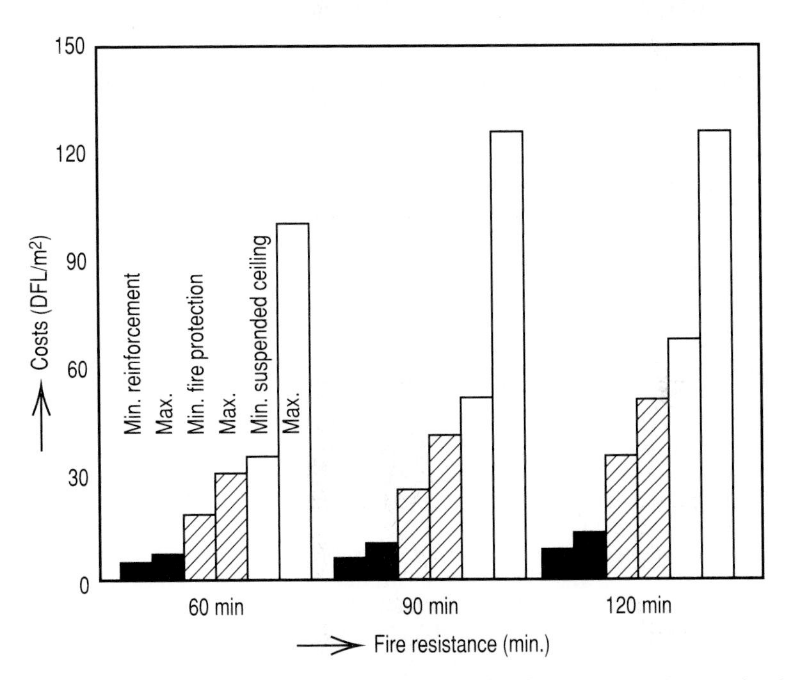

Fig. 3.24 Cost comparison for three methods of increasing fire resistance of composite slabs (cost per m^2, including attachment, in 1988 Dutch currency).

In the United States most composite floors have a 2-hr fire rating as determined by an ASTM standard fire test. Additional fire protection can be provided by using spray-on-type fireproofing or a suspended ceiling. In Europe fire resistance is normally achieved by using additional reinforcement in the troughs of end spans and by using the existing hogging reinforcement to control concrete cracking [in accordance with Eurocode 4 Part 10 (1990)]. The fire improvement methods for composite slabs are explained in detail hereafter.

1. *Suspended ceilings.* In many structures the slab is protected from fires from beneath by suspended ceilings. These may already be in place due to aesthetic, acoustic, or thermal requirements. This same ceiling may be used as a fire barrier.
2. *Spray-on insulation.* Insulating materials can be applied directly to the exposed decking. The fire rating obtained depends on the material chosen and the thickness of application. Spraying steel work is messy and some overspraying can be expected. It is important to protect expensive glazing, cladding panels, and other finishes during spraying. This may have implications for the construction schedule because cladding can be installed after spraying.
3. *Supplementary reinforcement.* Fire rating can normally be obtained by mesh reinforcement or reinforcing bars placed in the ribs of the decking. This is one method of increasing fire resistance. The advantage of placing additional positive-moment reinforcement in the ribs, over the decking itself, is that the surrounding concrete serves as insulation. When adding fire reinforcement in this manner, it is of utmost importance to ensure that the concrete cover is indicated on drawings.

The relative economics of using any of the foregoing fire protection methods differ among countries and over time. A cost comparison was performed in 1988 in the Netherlands and is illustrated in Fig. 3.24 (SIS, 1991). From this figure it can be seen that costs for additional reinforcement are the lowest. The method to increase the fire resistance should be chosen with respect to other criteria, such as acoustic and aesthetic requirements.

3.4 EXECUTION OF COMPOSITE SLABS

1 Work-Site Safety

Accidents on the job site are best prevented by first identifying dangerous activities and then implementing appropriate protective measures. It is not possible to indicate all possible dangerous job-site activities in this publication. Here only a few typical sources of danger connected with the use of lightweight decking are reviewed.

Many problems arise from a lack of information or poor communication between partners on the job site (job-site logistics). All relevant drawings and instructions should be circulated between all parties involved. Improving communication reduces not only accidents, but costly errors as well.

The following measures should be considered:

Decking should be fixed in place immediately after erecting the steel frame. For the workers the platform is an extra safety net.

Decking should be attached to the underlying frame as soon as possible. Decking and cutoffs are light and may become airborne in strong winds.

Depending on the type of profile, during construction workers should take care of seam laps that cannot be walked on.

At the end of the workday, decking bundles that were opened should be tied together.

2 Packing and Storage of Decking

European recommendations to avoid damage to decking prior to its placement are given in ECCS (1983) and adopted in SIS (1991).

Working with profiled-steel thin-walled sheet requires special attention due to the unique properties of the product. In general the decking has a self-weight of 60 to 100 MPa (8700 to 14,500 psi) and thus is very light. Because of the very small thickness of the decking, it can easily be damaged and blown away by the wind.

In general the decking will be delivered in bundles. The designer should try to specify all decking of the same length for a given project. For fabrication it is recommended to bundle decking having the same length. This could avoid unnecessary handling at the job site. For transport and storage it is important to handle the heavy bundles with care to avoid damage to the corners and edges. Attention should be given also to preventing movement of the decking.

Fig. 3.25 Some measures to avoid white rust on the decking (ECCS, 1983b).

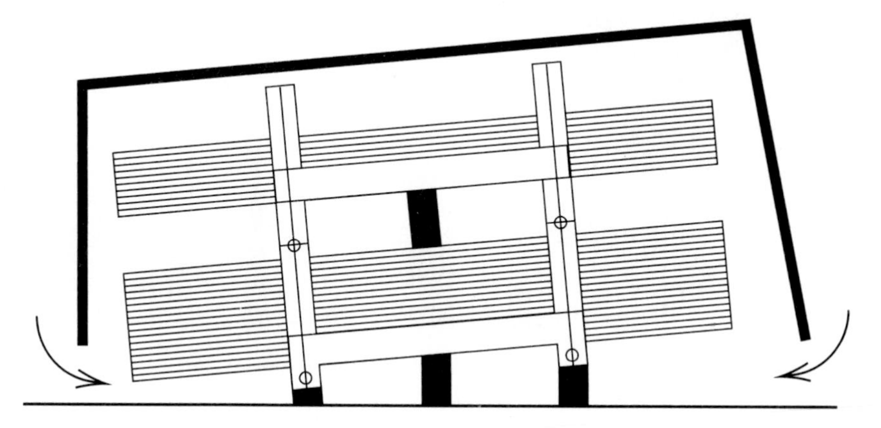

Fig. 3.26 Correct storage of panel bundles on site.

Decking should be delivered not more than 1 month in advance of its anticipated use. It should be stored well at the construction site. Bundles of decking that have to be stored for longer periods of time should be in a warehouse with low humidity and small temperature variations (Fig. 3.25). When stored outside for short periods of time, decking should be placed and covered as illustrated in Fig. 3.26. This ensures that it is not exposed directly to rain and sun. Condensation must be allowed to drain.

Decking bundles are normally stored in frames consisting of wooden blocks. These blocks allow lifting slings to be inserted (Fig. 3.27). Short lengths for decking (up to 3 m [10 ft]) may be lifted with sloping slings. Longer deckings should be lifted with a spreader beam. It is highly recommended to lift each bundle using these wooden blocks or other protective material and not to put lifting slings directly in contact with the deckings. For storage of several bundles, the wooden blocks should be placed on top of each other.

3 Overview of European Recommendations on Tolerances for Decking

Recommendations have been issued in Europe for several decking properties. The recommendations are contained in ECCS (1983) and, with respect to embossments, in Eurocode 4 Part 1 (1990). These recommendations were adopted in SIS (1991) and are briefly reviewed here.

Steel Core Thickness t. Tolerances on nominal steel core thickness must be in accordance with Euronorm 148 (1979). For decking in composite slabs Eurocode 4 Part 1 (1990) gives a core thickness tolerance of +10% and −5%.

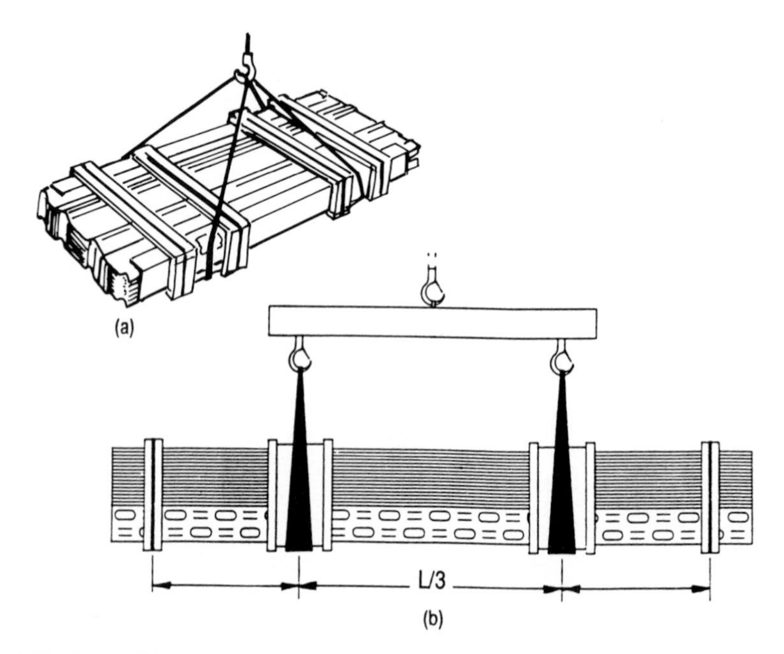

Fig. 3.27 Correct lifting arrangements for panel bundles. (*a*) Short panels. (*b*) Long panels.

Profile Height h. The allowable tolerances on the nominal profile height are −1% of the profile height and +4% of the profile height with a maximum of 2 mm (0.08 in.) (Fig. 3.28).

The depth of flange and web stiffeners may differ by a maximum of −10% of the nominal value given by the manufacturer.

Cover Width. The cover width must be measured at 200 mm (8.0 in.) from the decking ends. Allowable tolerances on the nominal cover width are (for both measured values):

For profiles ≤55 mm (2.2 in.)	1% of nominal cover width
For profiles >55 mm (2.2 in.)	2% of nominal cover width

The allowable difference between both measured widths of one decking length is:

For profile heights ≤55 mm (2.2 in.)	1.5% of nominal cover width
For profile heights >55 mm (2.2 in.) and ≤80 mm (3.2 in.)	2% of nominal cover width
For profile heights >80 mm (3.2 in.)	3% of nominal cover width

Deformations in Plane of Decking. The allowable tolerances with regard to the theoretical profile shape are $g \leq 0.002G$, with a maximum of 10 mm (0.4 in.) (Fig. 3.29).

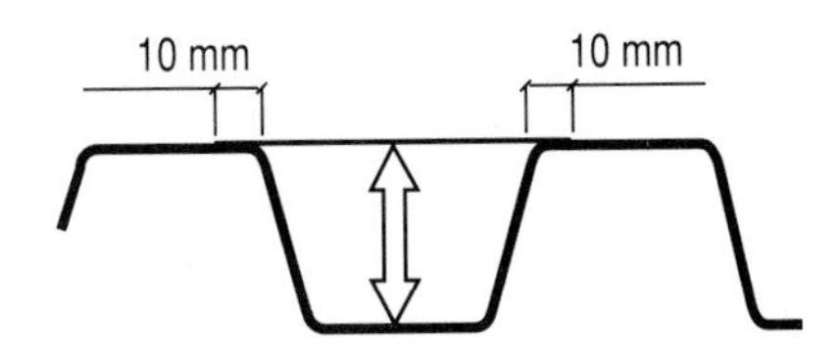

Fig. 3.28 Determination of profile height *h*.

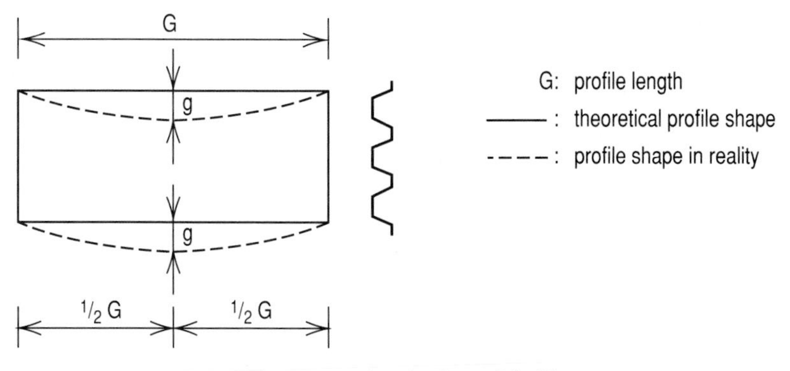

Fig. 3.29 Deformations in plane of decking.

Decking Length. The allowable tolerance on the nominal decking length is 0.3%, with a maximum of 20 mm (0.8 in.).

Radius. The allowable tolerances on the nominal inner radius between flanges and webs are ±2 mm (±0.08 in.). The radius should be measured 200 mm (8 in.) from the decking end.

Flatness of Decking. The allowable tolerance on the flatness of the decking is 3% of the nominal cover width. Determination of the tolerance should be made in the middle of a decking, placed on two supports at a distance of 5 m (16 ft). The decking must be pressed down on the supports (Fig. 3.30).

Embossments. Embossment dimensions may differ a maximum of −10% of the nominal value. The distance between the embossments may differ a maximum of ±5% of the nominal value.

The manufacturer of the decking should specify the shape and dimensions of the embossments in the tests. Because the longitudinal shear capacity of the composite floor depends on the shear connections between decking and concrete, the decking must be fabricated within previously mentioned tolerances.

4 Good Construction Practice

In the Netherlands, SIS recommendations have been established for good construction practice (SIS, 1991). These are reviewed in this section. All on-site cutting or openings must be made with scissors (Fig. 3.31). Tools such as circular saws are not recommended because sparks of hot steel can attach to the surface, becoming corrosion initiation points. When screw holes are drilled, the correct drill for the combination of decking thickness and fastener diameter must be used. The sharpness of the drill must be such that the hole diameters are within allowable tolerances.

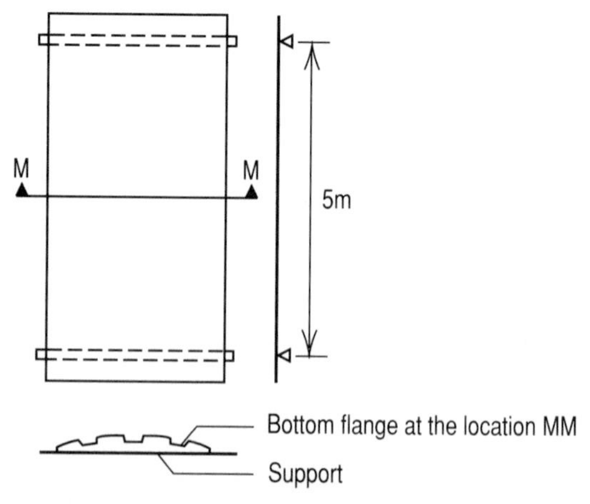

Fig. 3.30 Flatness of decking.

Before erection of the decking can start, the following is necessary:

The upper flanges of the steel beams on which studs will be welded are dry and free from rust or other elements that could have an unfavorable influence on the quality of the weld.

Props, if needed, are placed according to the designer's information. (A construction girder as indicated in Fig. 3.33 is also possible.)

The height of the bearings is measured correctly.

Good construction practice in both Europe and the United States suggests that loads be placed in such a way that the following conditions are satisfied:

The identification tag on each bundle should be checked against drawings to ensure that the correct geometry and thickness are used in the positions specified.

Bundles should be placed on the substructures with the overlapping seam oriented in the correct direction. This avoids having to turn each sheet manually before placement.

Installed decking should not be subjected to loads greater than those for which it was originally designed.

The substructure should not be subjected to loads greater than those for which it was originally designed.

Loads should be placed near columns or near supports, where possible.

Fig. 3.31 Use of scissors to cut decking.

Wooden blocking should be used to spread loads as evenly as possible.

Bundles should be lifted into place with care to avoid damaging individual decks. In general, slings should be placed near or on existing wooden shipping protections. Long sheets should be lifted using a lifting beam and two slings.

European recommendations (ECCS, 1983) indicate that workers should lift individual decks from bundles using the overlapping seams (Fig. 3.32). Once placed, decks should be connected as soon as possible. It is recommended to fasten a deck before installing the next one. After the deck is installed, all other necessary connections can be made. The decking should be fastened to the substructure at maximum intervals of 500 mm (20 in.). The seam lap connections between decks should also be fixed at maximum intervals of 500 mm (20 in.). In case the deck acts as a diaphragm, this distance should be reduced to 300 mm (12 in.) maximum.

5 Fastening

Decks must be attached to each other and to the underlying frame in order to be used safely as a working platform and as shuttering. The most common fastening technique between decking and a steel frame in the United States is arc puddle welding. This procedure burns a hole in the deck and fills the hole with weld material to fuse the deck to the structural steel (Heagler, 1987). Procedures and design data for welds are given by the American Welding Society (AWS) (1981) and AISI (1986). SDI (1987a) specifies the average spacing of welds in composite floor deck as 300 mm (12 in.). When beams are designed to act compositely, the shear studs welded through the deck into beams can replace the arc puddle welds as deck attachments (AISC, 1980). Often studs are not installed at the same time that the deck is erected, so some attachment is necessary to secure the deck safely until the studs are in place. Other fastenings to the frame can be powder-actuated or pneumatically driven pins or self-tapping screws. The use of mechanical fasteners is common practice in Europe. No unattached decks should be left at the end of a working day. In the United States typical seam lap fastening (connections between adjacent decks) is normally made using button

Fig. 3.32 Correct lifting of individual decks from a bundle (ECCS, 1983).

punches, tack welds, or screws (Heagler, 1987). Seam lap fastening of deck units (between supports) is important for the deck to be an effective working area. Button punching, when done with the usual crimping tool, is the least reliable. Welding needs metal-to-metal contact and usually leaves a hole in the deck—the edges of the holes are then connected. Self-drilling screws are the easiest to install. Before the designer specifies frame or seam connections, he or she should check the fire-rating design to see whether there are any connection restrictions.

6 Temporary Supports

Temporary supports (props) should be avoided whenever possible. Their use eliminates the major economic advantages of decking. In some cases, however, small floor areas may have to be propped. Such cases include infilled bays after crane removal and lift shafts which have nonstandard span lengths. As an alternative to the use of props, it is possible to use construction girders, which are placed on the bottom flange of the main girders of the steel frame (Fig. 3.33).

Good practice recommendations in the United Kingdom (Lawson et al., 1990), adopted by SIS (1991), call for the following provisions where using props. Props may be placed on the finished floor beneath, provided that the design

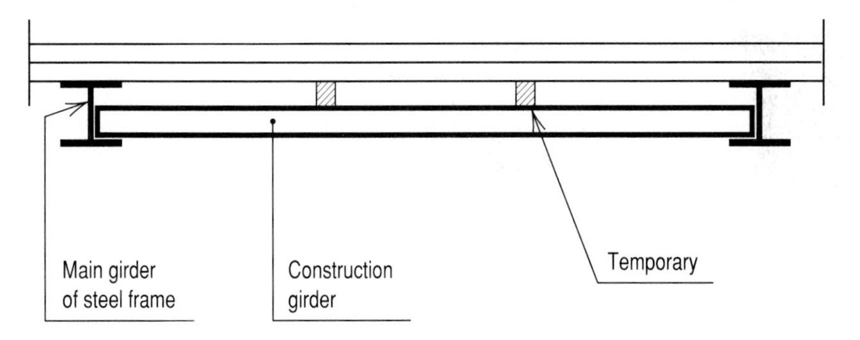

Fig. 3.33 Alternative to props for temporary support during construction.

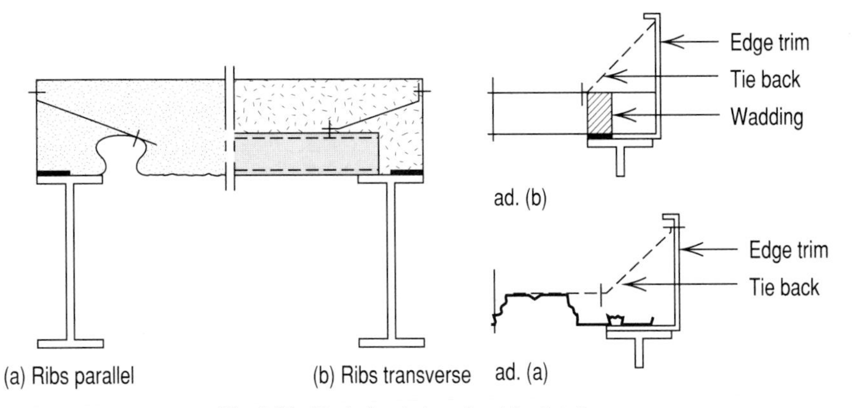

Fig. 3.34 Typical exterior edge trim details.

capacity of the floor exceeds the loads applied to the decking and concrete above [normally less than 4.5 kPa (94 psi)]. The supporting floor should have attained its design strength at this stage. If not, further props to the floor beneath are needed. Props should never be placed directly on the decking without additional props to the floor beneath. Props should not be removed until the floor has reached 75% of its design strength. This is normally achieved in 7 to 8 days.

The temporary props should be provided with wooden beams. The supporting width should not be less than 80 mm (3.2 in.). According to SIS (1991), the supporting width is normally approximately 120 mm (4.7 in.), which is achieved with two wooden beams of 65 × 165 mm (2.5 × 6.5 in.) side by side or one beam with another one turned on its side on top of it.

7 Preparation Prior to Placement of Concrete

Good practice recommendations in the United Kingdom (Lawson et al., 1990), adopted by SIS (1991), provide the following recommendations.

Edges of floors are often finished with thin-plate edge trim which is held in place by a tieback connected to the seam lap or end of the decking at a distance of 0.5 to 1 m (1.6 to 3.2 ft). The end of the decking is closed with a special wadding. Figures 3.34 and 3.35 give some examples.

At the free edge of a deck where the ribs are running parallel to the side a support (which could be temporary) will often be necessary. (See Fig. 3.36 for construction details.)

To avoid concrete spillage, gaps larger than 5 mm (0.2 in.) must be covered. Decking cutoffs can be used for this purpose. This is usually the case around columns. Gaps between butted decks can be taped (Fig. 3.37). The remaining decking cutoffs must be collected and disposed of in skips, which are placed above a beam or a column.

If the decking is installed by a subcontractor, as is usually the case in Europe, it is advised to make a formal agreement about who has to clean up the deck prior to concreting. Screws, stickers, plastics, rivets, and such must be removed.

Before concreting, parts of the anchor bolts that will not be encased in the concrete must be oiled. The pouring scheme must be in accordance with the

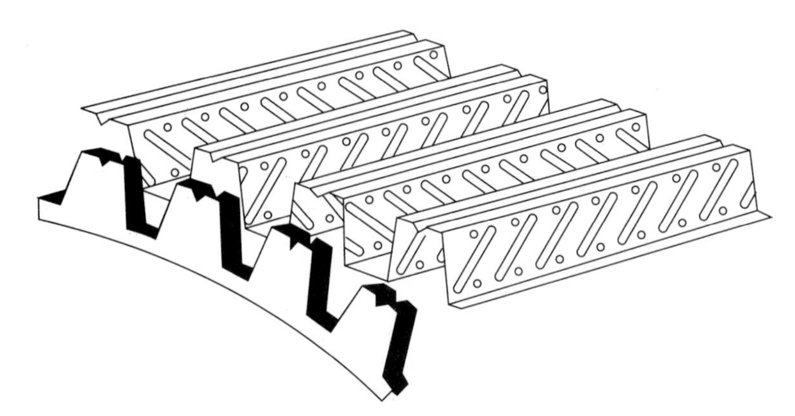

Fig. 3.35 Wadding placed underneath ribs before edge trim installation.

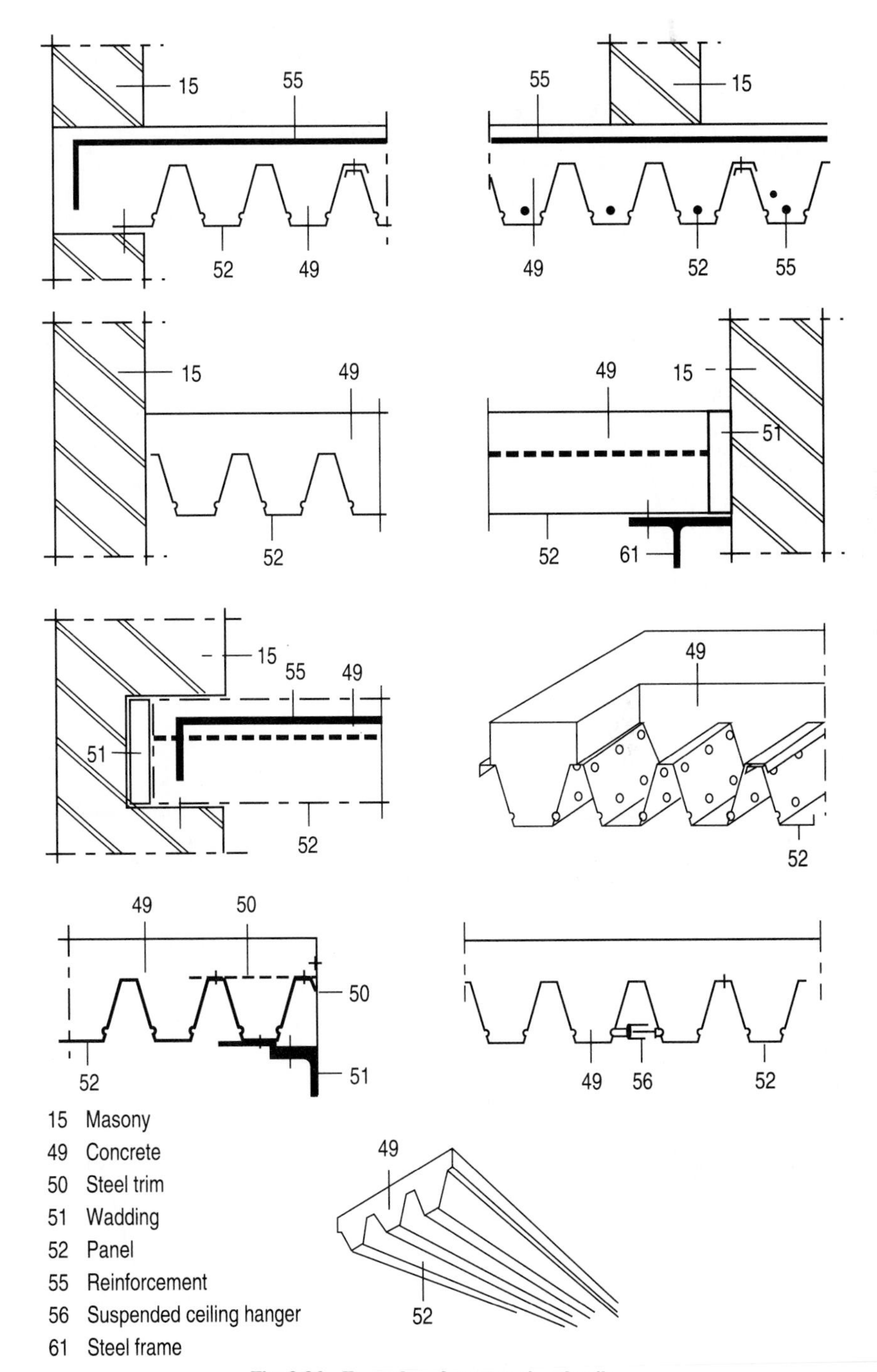

Fig. 3.36 Examples of construction details.

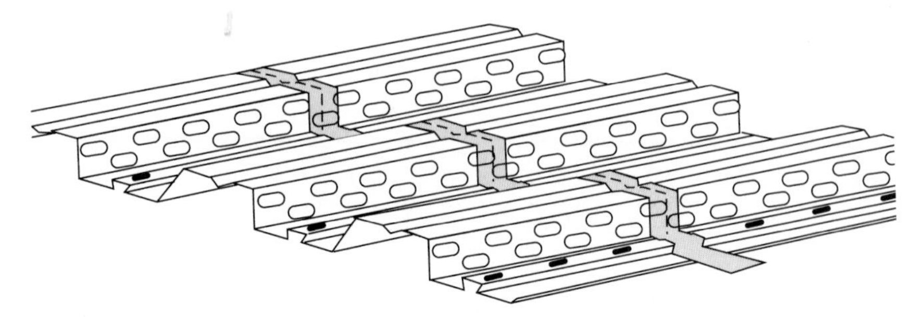

Fig. 3.37 Proper taping of butted panel ends.

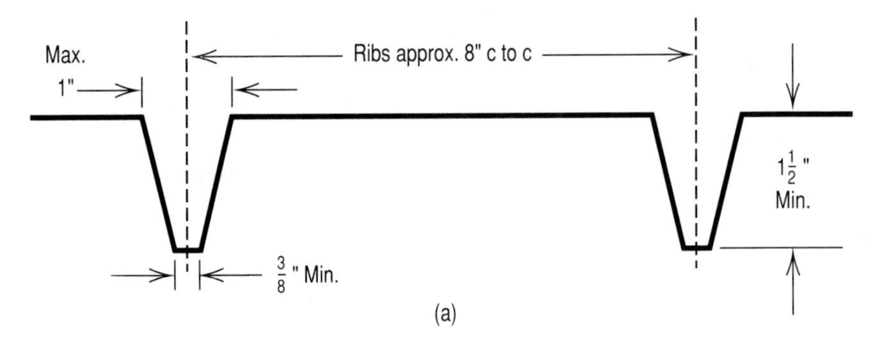

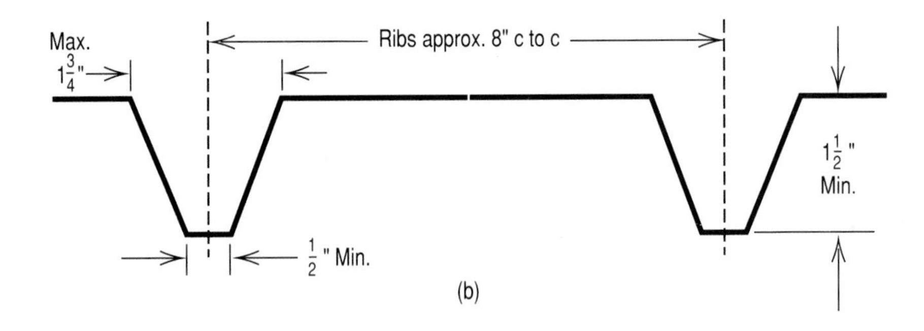

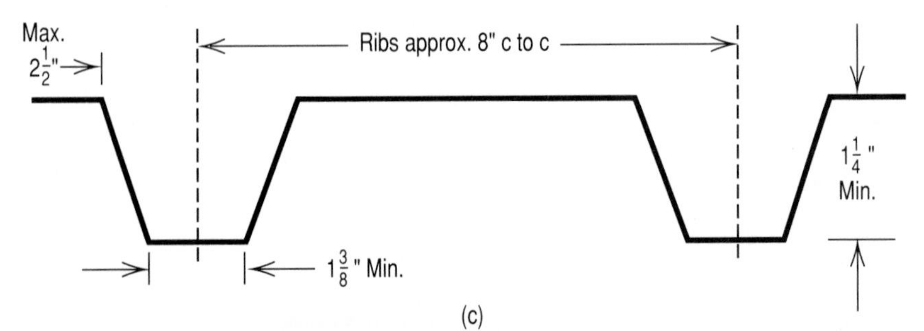

Fig. 3.38 Typical roof decking profiles. (*a*) Narrow-rib deck type NR. (*b*) Intermediate-rib deck type IR. (*c*) Wide-rib deck type WR.

loading considered in the design of the decking. Debonding between steel and concrete of already poured and partly hardened areas of the floor must be avoided.

3.5 ROOF DECKING

In this section constructional details unique to roof decking are examined. Other considerations, not discussed in this section, are similar to those presented earlier. The U.S. practice is reported in Heagler (1991) and SDI (1987c).

Typical roof decking in the United States is shown in Fig. 3.38. These decks are usually part of a layered "flat" roof system; they are not exposed roofing. Rib openings have been set to accommodate the spanning capabilities of different insulation boards. In the past, before energy costs were so crucial, the narrow-rib decks were sometimes used with very thin insulation boards. Now insulation boards are thicker, at least 50 mm (2 in.), and the wide-rib deck profile, which is structurally much more efficient, is the deck of choice. SDI has adopted 228 MPa (33 ksi) as the minimum material yield point for roof decking, and most load tables are based on this limit. In reality the yield point is seldom below 275 MPa (40 ksi).

1 Corrosion Protection

A key consideration for all deck products is the required service life and the environment in which a decking product must exist. For instance, if a roof has insulation board assembled with fasteners that penetrate the decking, then attention should be given to the corrosion of the fasteners and the decking. If water comes through a leaking roof membrane or if water continuously condenses underneath, then the decking should be protected; galvanizing is cost-effective protection. Humid areas or areas exposed to water should be protected against corrosion. Insurance requirements and fire rating can also affect the finish selection.

2 Openings

In this subsection Heagler (SDI, 1987c) provides a review of U.S. and Canadian provisions for roof deck openings. For most 38-mm (1.5-in.) roof decking the loss of one rib, by either denting or penetration, can be tolerated and no reinforcement will be required. The Canadian standard for steel roof deck (CSSBI 10M-86) allows an unreinforced opening of 150 mm (6 in.) as long as not more than two ribs are removed (Canadian Sheet Steel Building Institute, 1986). A simple and conservative estimate of the deck capacity would be to take out one rib (about 20% at the worst) and apply this to the uniform loads shown in the tables; in most cases the load capacity is greater than required anyway. A 150-mm (6-in.)-diameter hole will probably not bother the diaphragm strength; a dent can be larger than 150 mm (6 in.) and still carry the necessary horizontal load. Covering the dent or a 200-mm (8-in.) maximum hole with a 1.14-mm (0.045-in.) plate and carrying the plate to adjacent ribs could eliminate worries about insulation board spanning the dent and about a "soft spot" in the roof. For holes or dents greater than a rib, that is, 200 to 330 mm (8 to 13 in.), it would be advisable to use a 1.45-mm (0.057-in.) minimum plate. There are exceptions to this advice:

1. The hole may be located such that the deck can safely cantilever from each adjacent support.
2. A group of holes may be so close together that a frame is required.

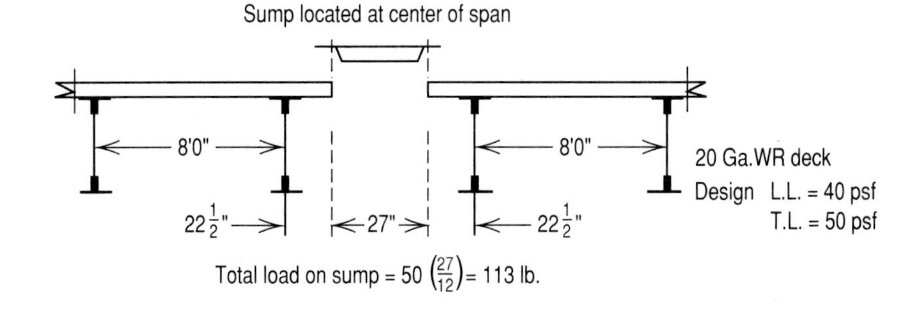

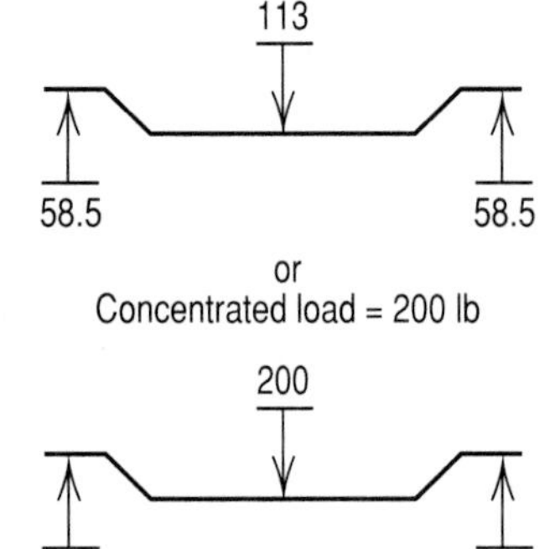

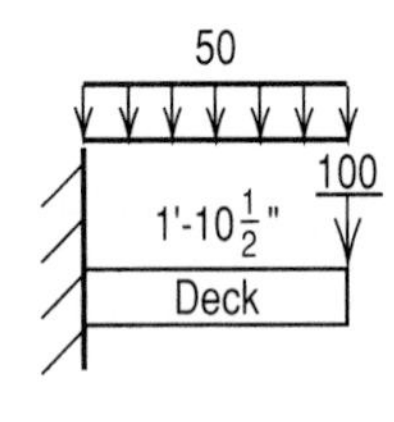

$$M = \frac{50\left(\frac{22.5}{12}\right)^2 (12)}{2} + 100\,(22.5) = 3305 \text{ in. lb}$$

Typical Sn = 0.184 of $1\frac{1}{2}$" deck

$$fb = \frac{3305}{0.184} = 18000 \text{ psi} < 20000 \text{ O.K.}$$

< 26600 O.K.

(Temp. man load)

Note:

Cantilever table in SDI Manual says 1'-10 $\frac{1}{2}$" O.K.
No reinforcing required.

Fig. 3.39 Sump pans; Example 1.

A special case of roof penetration is the sump pan. Generally, when properly attached the sump pan will carry the load of the deck it replaces. It also acts as a small header to transfer loads into adjacent uncut decking. Approximate per foot (of width) section properties of a standard [1.9-mm (0.075-in.)] sump pan are $I = 1.5 \times 105\ \text{mm}^4$ (0.36 in.^{4}), $SP = 0.33 \times 10^4\ \text{mm}^3$ (0.20 in.^{3}). Sump-pan analysis methods are shown in Figs. 3.39 and 3.40 as a reinforcing technique.

3 Fastening

Heagler (SDI, 1987c) provides the following commentary on U.S. construction practices. Burn holes in seam laps, caused by welded seam-lap attachments, are spaced far enough apart not to cause problems. However, most deck manufacturers advise not to weld the seam laps of 7.1-mm (0.028-in.) deck or thinner because of the difficulty of obtaining a good weld. Typically a seam-lap weld will cause a hole; the perimeter of the weld does the work. Burn holes near intermediate supports are unlikely to cause much loss of strength. (These holes are usually caused by the welder searching for the unseen structural member.)

Distributed small dents, such as those caused by foot traffic, will not cause a structural problem. But if the denting covers a large percentage of the job, the insulation board may be better attached with mechanical fasteners rather than by adhesives. Holes caused by mechanical fasteners or screws are no problem. If the deck is not galvanized, it is good practice to paint burned or abraded areas.

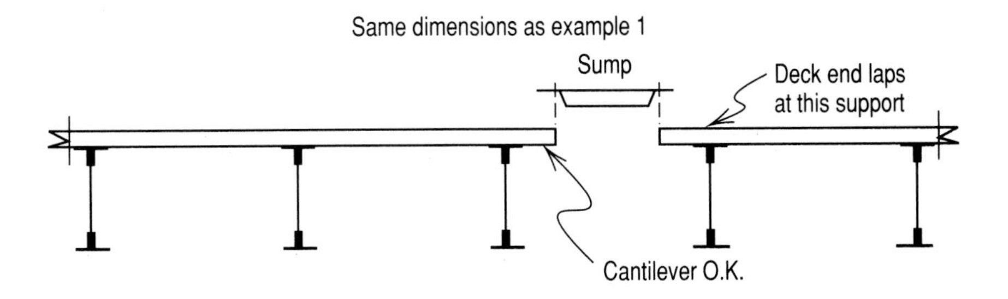

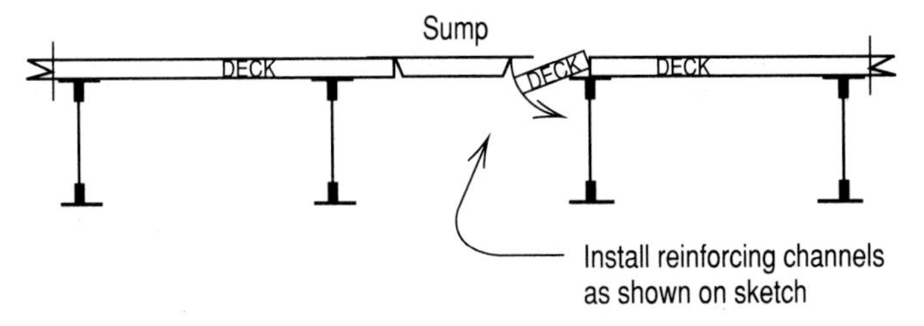

Fig. 3.40 Sump pans; Example 2.

4 Roof Design

European roof design is governed by Eurocode 3 Annex A (1991). These calculations must be performed to ensure adequate strength and stiffness. Another European document (ECCS, 1983) is concerned with good roofing practices. Heagler (SDI, 1987c) provides the following commentary on roof design practices in the United States.

The design of roof decking obviously must take into account vertical gravity loads and horizontal loads from wind or earthquake. Less obvious but very important are fire-rating requirements, uplift wind loads, and maintenance (or construction) considerations. Vertical gravity loads are usually uniformly distributed loads caused by snow or, rarely, water ponding. Snow drifting must be considered. SDI provides generic uniform load tables, and each manufacturer also shows load tables in catalogs. Since snow drifting normally results in a

Table 3.6 Uniform total (dead and live) load (per ft^2) for type B—wide-rib roof decking (based on SDI criteria)

Number of spans	Gage	Span 5 ft-0 in.	5 ft-6 in.	6 ft-0 in.	6 ft-6 in.	7 ft-0 in.
1	22	99	77	61	50	42
1	20	122	94	75	61	51
1	18	173	132	104	84	69
1	16	220	168	132	106	87
2	22	114	94	79	67	58
2	20	141	117	98	84	72
2	18	189	156	131	112	97
2	16	235	194	163	139	120
≥3	22	142	117	99	84	71
≥3	20	177	146	123	105	87
≥3	18	237	196	164	140	121
≥3	16	294	243	204	174	150

Number of spans	Gage	Span 7 ft-6 in.	8 ft-0 in.	8 ft-6 in.	9 ft-0 in.	9 ft-6 in.	10 ft-0 in.
1	22	36	32	28	25	23	21
1	20	43	37	33	29	26	24
1	18	58	50	43	38	34	30
1	16	72	61	53	46	41	36
2	22	50	44	39	35	31	28
2	20	63	55	49	44	39	35
2	18	84	74	66	58	52	47
2	16	105	92	81	73	65	59
≥3	22	60	51	44	39	34	31
≥3	20	72	61	53	46	41	36
≥3	18	101	85	72	63	55	48
≥3	16	127	107	91	78	68	59

nonuniform (sloping) load, it may require special analysis. The uniform load capacity of most decking is greater than the applied loads. A look at the tables (see Table 3.6) is usually enough to select the decking. Uplift loads can also be handled easily by the decking. The important check to make is on the fastener capacity. However, up to now there have been no reports of an uplift failure where decking has separated from the framing steel or bar joists.

Maintenance (construction) loads are human loads. The SDI uses an 890-N (200-lb) concentrated load, distributed over a width of 0.3 m (1 ft), located at decking midspan. With this loading the bending stress in the decking is limited to 179 MPa (26 ksi), and the deflection is limited to $^1/_{240}$ of the span length. For most roof decking this rule provides a span-length limit. The Factory Mutual Corporation (FMC) has a different concentrated loading, which can result in a somewhat shorter maximum span than found by the SDI method. The SDI publishes maximum allowable deck spans based on maintenance loading. Table 3.7 shows one manufacturer's published maximum span table for 38-mm (1.5-in.) wide-rib roof deck.

The design for vertical loading can mostly be done by consulting appropriate tables. Even snow drifting can usually be checked simply by calculating the worst resulting load and verifying that the decking is sufficient. Horizontal loading from wind or earthquake requires much more analysis (see also Chapter 4). However, the wind or earthquake analysis results in a load, or series of loads, being applied at the edge of the deck along the perimeter steel. The conventional model is to consider the whole roof structure as a large plate girder. The roof deck is then the web (Fig. 3.41).

A suggested method of analysis is to draw a shear diagram of the roof plate as it is loaded at the edges by the horizontal forces. The area under the shear diagram (divided by the depth of the plate times the stiffness) is the shear deflection (Fig. 3.42). Chapter 4, Section 4.2, gives a detailed treatment of this method.

SDI provides extensive tables for checking the strength and stiffness of the deck. Different fastener types, patterns, and combinations are covered. Figure 3.43 provides an example.

The diaphragm strength and stiffness are greatly influenced by the amount and type of fastening. Because fastening is also a cost consideration, it may be advisable to zone the roof with a fastening schedule: heavy connections in the high-shear areas and fewer connections in areas of less shear. The designer must check for loads coming in other directions. Roof decking design can then be

Table 3.7 Maximum recommended spans for roof applications for 38-mm (1.5-in.) type B—wide-rib roof decking based on SDI criteria

Metal thickness, in.	Gage	Single span	Multiple spans	Cantilever span
0.0295	22	5 ft 10 in.	6 ft 11 in.	2 ft 0 in.
0.0358	20	6 ft 7 in.	7 ft 9 in.	2 ft 4 in.
0.0474	18	7 ft 11 in.	9 ft 5 in.	2 ft 10 in.
0.0598	16	9 ft 0 in.	10 ft 8 in.	3 ft 0 in.

summarized as follows:

1. Select decking gage and profile based on vertical loading.
2. Check that maximum recommended span is not exceeded.
3. Choose fastening based on diaphragm needs.
4. Check uplift on fasteners.

Some final points to remember on roof deck diaphragms are:

1. The roof decking is not considered to be "reinforced" by adhered insulation board or to act compositely with it.
2. The traditional one-third stress increase allowed for wind or earthquake loads should not be applied to the loads shown in SDI tables.
3. The loads (and stiffnesses) shown in the SDI tables apply to the decking regardless of its orientation (that is, the table values apply parallel and perpendicular to the ribs).
4. The designer must determine the amount of allowable deflection in the diaphragm.

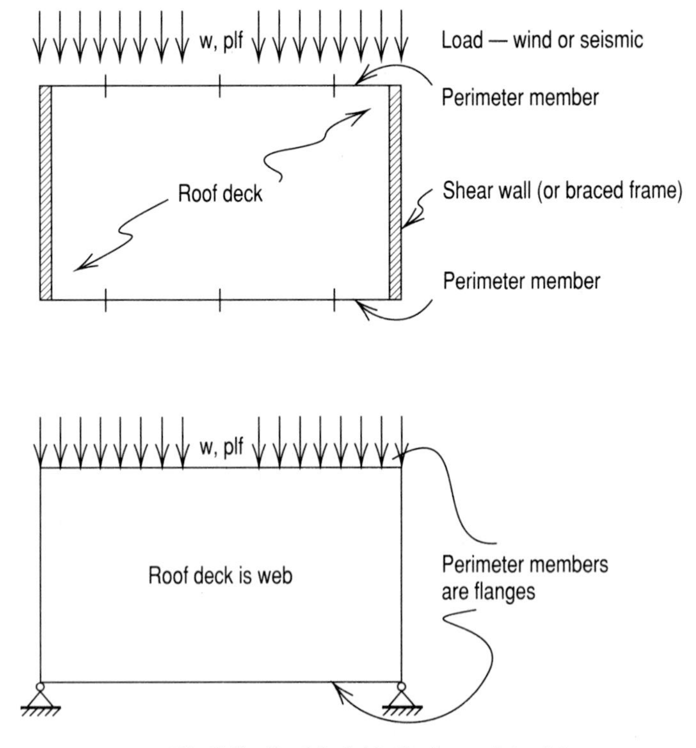

Fig. 3.41 Roof deck idealized as a plate girder.

3.6 CONCLUSIONS FOR ROOF AND FLOOR DECKING

In both the United States and Europe, due to the available number of design guides and tables, the selection of a floor or roof decking system is relatively easy. In most cases enough information is available to design for both vertical and in-plane loading. The designer should be aware of site conditions, both during and after construction. An appropriate finish coating on the decking must be chosen with care.

3.7 SANDWICH PANELS

Sandwich construction (Allen, 1969) is a composite panel consisting of two thin outer sheets (facings) of high-density material and a thick interior layer (core) of low-density material. Sandwich construction has been widely applied in aircraft and structural engineering since before World War II. The structural

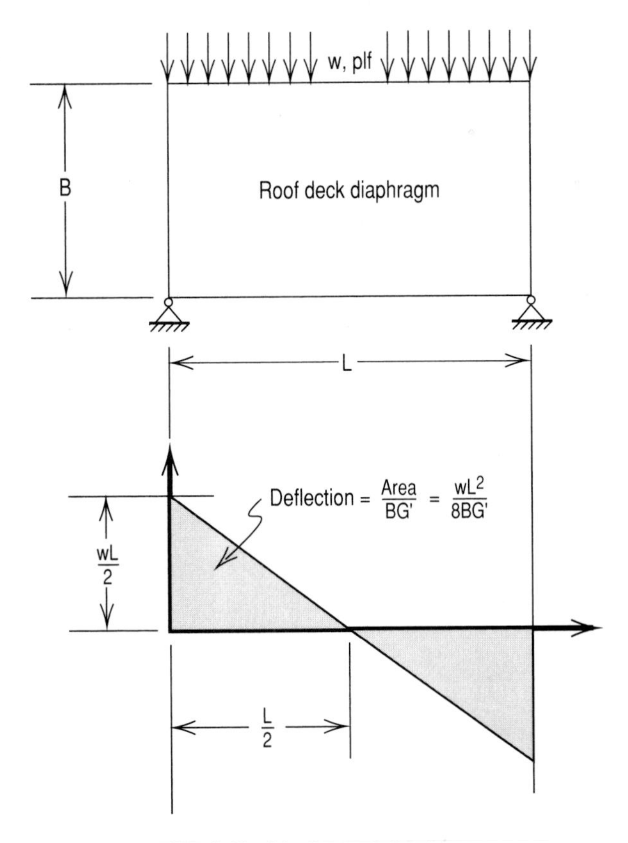

Fig. 3.42 Diaphragm deflection.

analysis of sandwich panels with thin flat facings has been investigated as early as the 1940s, particularly for aeronautical applications (Allen, 1969; Gough et al., 1940; Hoff and Mautner, 1956). However, research and development of architectural sandwich panels with formed facings (Fig. 3.44) began only in the early 1970s, pioneered by Chong and his associates. These panels are popular due to their superior structural efficiency, mass productivity, insulation qualities, transportability, fast erectability, prefabricatability, durability,

v14 Standard 1.5" deck

Frame fastening: 5/8" welds on 30/4 pattern.
Stitch fastening: #10 screws (buildex) Safety factor: 2.75

t = design thickness = .0295"
Design shear, plf

Stitch connectors per span	Span, ft. 3.0	3.5	4.0	4.5	5.0	5.5	6.0	6.5	7.0	K1
0	340	300	270	240	215	195	175	160	150	0.728
1	395	350	315	285	260	235	215	195	180	0.536
2	440	395	355	325	295	275	255	230	215	0.424
3	475	430	395	360	330	305	285	265	245	0.350
4	510	465	425	395	365	335	315	295	275	0.299
5	540	495	460	425	395	365	340	320	300	0.260
6	565	525	485	450	420	395	370	345	325	0.231

$D_{wr} = 1377$ $D_{ir} = 1547$ $D_{nr} = 1608$ K2 = 870
Substitute these values into the equation for G' as appropriate.

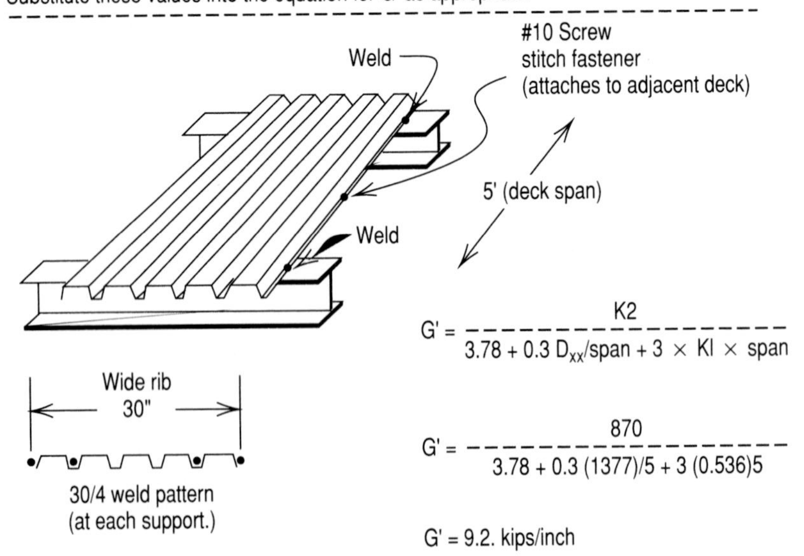

Fig. 3.43 SDI diaphragm table.

and reusability. The formed facings are for architectural appearance and structural stiffness.

Chong and Hartsock (1972, 1974) presented a method to predict the localized wrinkling instability of such panels. Subsequently Chong and his associates have investigated flexural behavior (Chong et al., 1979; Hartsock and Chong, 1976; Tham et al., 1982) and thermal stresses (Chong et al., 1976, 1977, 1982b) for both simple and continuous spans, axial stability (Tham et al., 1982; Chong et al., 1984; Cheung et al., 1982), and vibration (Chong et al., 1982a). Some of these findings were summarized in a keynote address (Chong, 1986). The formed or flat-faced sandwich panels can be used as bearing walls, curtain walls, partitions, exterior enclosures, and such. In this chapter the structural behavior of formed and flat-faced sandwich panels, including wrinkling, flexural stresses and deflections, axial stability, thermal stresses, and vibration, is presented, summarizing more than a decade of research. Methods used include analytical, numerical, and experimental procedures. Results by different methods are compared. The

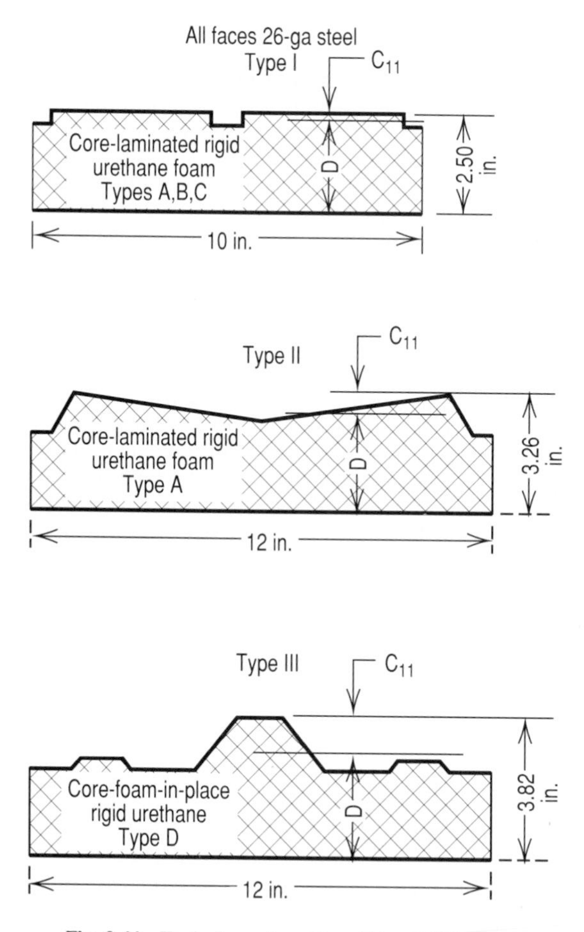

Fig. 3.44 Typical panel profiles. (1 in. = 25.4 mm.)

classical references for flat-faced sandwich panels are given by Allen (1969) and Kuenzi (1960); ASCE conducts regular literature surveys (1969), including sandwich panels under the composite construction heading.

In Europe ECCS committee 7 has drafted two recommendations:

1. A recommendation with design rules and testing procedures for sandwich panels (ECCS, 1991)
2. A recommendation which provides information concerning good practice and recommended details for sandwich panels (ECCS, 1990)

The International Association for Bridge and Structural Engineering (IABSE) colloquium in Stockholm in 1986 presented three papers concerning sandwich panels (Chong, 1986; Höglund, 1986; Hassinen, 1986), as did the IABSE symposium in Brussels in 1990 (Davies and Hakmi, 1990; Berner and Stemmann, 1990; Hassinen and Helenius, 1990).

1 *Wrinkling Instability*

The critical stress in the compression face is first evidenced by localized wrinkling, in which the wavelength is of the same order as the core thickness. Normally the exterior faces are formed to improve appearance and to resist buckling and "oil canning." There are numerous theories dealing with wrinkling in axially loaded sandwich columns. However, there are few references on the flexural wrinkling of panels with flat faces (Allen, 1969; Gough et al., 1940; Hoff and Mautner, 1956; Harris and Nordby, 1969) or on the commercially significant case of panels with formed faces (Chong and Hartsock, 1972, 1974).

In this chapter five types of panels were subjected to beam tests with center point loading. Three face configurations of 26-gage or 0.5-mm (0.02-in.) steel (Fig. 3.44) and four kinds of rigid urethane foam were tested. Strain gages were mounted at the points of maximum compressive stress. The experimental critical stresses were compared to those derived from a mathematical analysis wherein the compressed face is considered as a beam supported by the continuous elastic medium of the core, which provides an equivalent spring constant. The key to the solution of the problem is choosing four realistic boundary conditions for the core and the introduction of an equivalent face thickness. The spring constant is a complicated function composed of the wavelength of the buckled shape and the properties of the core. Once the constant has been found, it is substituted into the equation for the beam on an elastic medium, which is then solved for the minimum critical compressive force. The critical stress can be computed from the properties of the face and core. The calculation is greatly simplified by a family of curves (Chong and Hartsock, 1974). It was observed that experimental and theoretical critical wrinkling stresses are in close agreement.

2 *Flexural Behavior*

The flexural behavior of sandwich panels subject to bending was investigated analytically (Hartsock and Chong, 1976; Hartsock, 1969), experimentally (Hartsock and Chong, 1976), and numerically (Tham et al., 1982). The basic assumptions were that (1) the Young's moduli of the faces are large compared to those of the core, and (2) adequate adhesion exists between the faces and the core. The total moment is equal to the sum of the moment due to composite action and the moment due to bending of the faces about their own centroidal axes.

The two test panel profiles (types I and III) are shown in Fig. 3.44. Strain gages were mounted at the highest points of the profiles at one-sixth, one-third, and midspan. At midspan, gages were also located at the bottom of the grooves of the laminated panels and at different elevations of the foam-in-place panel. The panels were tested in a suction box, which was covered with a 0.15-mm (0.006-in.) polyethylene film taped to the sides of the box. Deflection was measured by means of dial indicators located over one support and at one-sixth, one-third, and midspan. Suction was applied by a vacuum pump through a hole in one side of the box and controlled by means of a small damper. Pressure was measured with a water manometer.

Computed deflections were in good agreement with the experimental values. The moduli were reasonable, and the shapes of the computed deflection/span curves were in good agreement with experiment. In conclusion, within the buckling loads, the methods outlined herein provide a reasonably accurate method of calculating the deflections and stresses in sandwich panels with formed faces. There is good agreement between experimental and theoretical values. The theory is applicable to combinations of formed and flat faces, and to faces of different or the same material, subjected to flexural loading.

Numerically the finite-strip method (Cheung, 1968a, 1968b, 1976a) and the finite-prism method (Cheung et al., 1976b) are especially efficient in analyzing prismatic members such as these architectural sandwich panels, whereas it would be prohibitive in cost and time to apply the finite-element method (Zienkiewicz and Cheung, 1967) to these panels. In this chapter the finite-prism method, in combination with the finite-strip method, was applied to the sandwich panels with formed facings. The results compared closely with those obtained analytically and experimentally by Hartsock and Chong (1976) for flexural bending.

3 Axial Stability

The buckling of sandwich panels due to axial compression was investigated using the finite-strip-prism model (Tham et al., 1982). The stability analysis of the system is equivalent to finding a set of nontrivial solutions to the set of homogeneous equations of a generalized eigenvalue problem. Solutions compared favorably with published data.

Due to uneven curing or other reasons, the sandwich cores may have variable stiffnesses throughout their thickness. Sandwich panels with flat faces and variable core stiffnesses were investigated by the finite-layer method (Chong et al., 1984; Cheung, 1976a). The faces were assumed to be isotropic and equal in thickness. The stiffness of the core follows the distribution assumption:

$$E_c t_c = \sum E_i t_i \tag{3.2}$$

where E_c and t_c are the mean Young's modulus and the thickness of the core, and E_i and t_i are the Young's modulus and the thickness of each layer of the core, respectively. The buckling loads of a sandwich square plate with homogeneous core material were used for confirmation and comparison, and close correlation with published data was observed.

For variable stiffness of the core of sandwich plates, optimum results occur when the stiffness ratios of the outer layers to the inner layers are equal to 1 (Chong et al., 1984). This means that the best results are obtained when the core material is homogeneous throughout the thickness of the core. Thus it is ideal to have a homogeneous core in any fabrication process. If the core is cured improperly, the outer layers tend to be different in stiffness compared with the inner layers. A strong outer layer or strong inner layer will reduce the strength of the plate. This section presented quantitative data on the buckling strength due to uneven core stiffnesses. Qualitatively it can be seen that if any part of the core is weakened or strengthened at the expense of the overall core stiffness, then the buckling strength is lowered. Intuitively, efficient sandwich columns should have stiff faces and weak cores such that the radii of gyration (by the transformed-area concept) are maximized.

4 Free Vibration

Free vibration of formed-face sandwich panels was analyzed by finite-prism-strip methods (Chong et al., 1982a). The mode shapes (eigenvectors) derived can be used in model analysis for forced vibrations due to random excitation and the like. The consistent mass matrix of the bending strip can be easily expressed in terms of the shape function $[N]$ and the basic unit submatrix as:

$$ {}^{s}[M_{ij}]^{e}_{mn} = \int \rho_f t_f [N_i]^T_m [N_j]_n d(\text{area}) \tag{3.3} $$

in which ${}^{s}[M_{ij}]^{e}_{mn}$ is the basic unit submatrix of the consistent mass matrix, ρ_f is the density of the thin face and t_f is the face thickness. With proper choice of harmonic series, the stiffness matrices are zero for unequal m and n, and they are decoupled. Hence the free-vibration equation can be written as

$$ -\omega^2\, {}^{s}M^{p}_{m}\, {}^{s}\bar{\delta}^{p}_{m} - \omega^2\, {}^{s}M^{B}_{m}\, {}^{s}\bar{\delta}^{B}_{m} - \omega^2\, {}^{p}M_{m}\, {}^{p}\bar{\delta}_{m} + {}^{s}K^{p}_{m}\, {}^{s}\bar{\delta}^{p}_{m} + {}^{s}K^{B}_{m}\, {}^{s}\bar{\delta}^{B}_{m} + {}^{p}K_{m}\, {}^{p}\bar{\delta}_{m} = 0 \tag{3.4} $$

where ${}^{s}M^{p}_{m}$ is the in-plane mass matrix of the bending strip, ${}^{s}M^{B}_{m}$ is the bending mass matrix of the bending strip, ${}^{p}M_{m}$ is the mass matrix of the finite prism, ${}^{s}K^{p}_{m}$ is the in-plane stiffness matrix of the bending strip, ${}^{s}K^{B}_{m}$ is the bending stiffness matrix of the bending strip, ${}^{s}K_{m}$ is the stiffness matrix of the finite prism, ${}^{s}\bar{\delta}^{p}_{n}$ and ${}^{s}\bar{\delta}^{B}_{m}$ are the in-plane and bending interpolating parameters of the bending strip respectively, and ${}^{p}\bar{\delta}_{m}$ are the interpolating parameters of the finite prism. Equation 3.4 is an eigenvalue problem with ω^2 as the eigenvalue, and all the variables have been defined (Chong et al., 1982a).

Numerically the present method is as accurate as published data for flat sandwich plates (Ahmed, 1971; Academia Sinica, 1977). The absolute error in both cases is less than 3% for the five modes. However, the present method is more generally capable of dealing with a wide variety of sandwich panels with or without formed faces.

5 Thermal Stresses and Deflections

Due to the superior insulation quality of sandwich panels, they have been used in places of extreme climates. It is not unusual that the temperature difference between the inside and outside wall surfaces may exceed 55.6°C (100°F). Due to the flexural rigidity of cold-formed facings, thermal stresses are present even in simple span conditions (Chong et al., 1976). A series of experiments was conducted to study the stresses and deflections induced when a sandwich panel is exposed to a temperature gradient between the two faces. The sandwich panels were tested for both single- and two-span conditions, with the temperature difference between the two faces reaching up to 55.6°C (100°F).

Formulated as an ordinary fourth-order differential equation with suitable boundary conditions, theoretical expressions were derived for deflection, flexural stresses in the facings, and shear stresses in the core. Numerically the finite-prism-strip method was used (Chong et al., 1982b). Experimental data, numerical analysis, and theoretical predictions were found in reasonable agreement.

6 Conclusions

Structural behavior of architectural sandwich panels was investigated analytically, experimentally, and numerically. Close agreements among these independent methods show that the results are reliable. Since experiments usually are very time-consuming, theoretical (analytical) or numerical analyses, or both, are preferred for the design and optimization of such panels. If necessary, confirmation tests can be performed.

3.8 CONDENSED REFERENCES / BIBLIOGRAPHY

Academia Sinica 1977, *Bending, Stability and Vibration of Sandwich Panel*

Ahmed 1971, *Free Vibration of Curved Sandwich Panels by the Method of Finite Elements*

AISC 1980, *Manual of Steel Construction*

AISI 1986, *Specification for the Design of Cold-Formed Steel Structural Members*

Allen 1969, *Analysis and Design of Structural Sandwich Panels*

ASCE 1969, *Literature Survey of Cold-Formed Structures*

ASCE 1984, *Specification for the Design and Construction of Composite Slabs and Commentary on Specifications for the Design and Construction of Composite Slabs*

AWS 1981, *Structural Welding Code, Sheet Steel*

Berner 1990, *Sandwich-Panels with Steel Facings and Different Core Materials*

Canadian Sheet Steel Building Institute 1986, *Standard for Steel Roof Deck*

Cheung 1968a, *Finite Strip Method in the Analysis of Elastic Plates with Two Opposite Simply Supported Ends*

Cheung 1968b, *Finite Strip Method Analysis of Elastic Slabs*

Cheung 1976a, *Finite Strip Method in Structural Analysis*

Cheung 1976b, *Three-Dimensional Analysis of Flexible Pavements with Special Reference to Edge Loads*

Cheung 1981, *Finite Strip Method in Structural and Continuum Mechanics*

Cheung 1982, *Buckling of Sandwich Plate by Finite Layer Method*

Chong 1972, *Flexural Wrinkling Mode of Elastic Buckling in Sandwich Panels*

Chong 1974, *Flexural Wrinkling in Foam-Filled Sandwich Panels*

Chong 1976, *Thermal Stress in Determinate and Indeterminate Sandwich Panels with Formed Facings*

Chong 1977, *Thermal Stress and Deflection of Sandwich Panels*

Chong 1979, *Analysis of Continuous Sandwich Panels in Building System*

Chong 1982a, *Free Vibration of Foamed Sandwich Panel*

Chong 1982b, *Thermal Behavior of Foamed Sandwich Plate by Finite-Prism-Strip Method*

Chong 1984, *Analysis of Thin-Walled Structures by Finite Strip and Finite Layer Methods*

Chong 1986, *Sandwich Panels with Cold-Formed Thin Facings*

Council on Tall Buildings 1983, *Developments in Tall Buildings 1983*

Council on Tall Buildings 1979, *Cold-Formed Steel*

Council on Tall Buildings Group SB, 1979, *Structural Design of Tall Steel Buildings*

Daniels 1990, *Bearing Capacity of Composite Slabs: Mathematical Modeling and Experimental Study*

Davies 1990, *Local Buckling of Profiled Sandwich Plates*

ECCS 1983, *European Recommendations for Good Practice in Steel Cladding and Decking*

ECCS 1990, *European Recommendations for Sandwich Panels, Part II: Good Practice*

ECCS 1991, *European Recommendations for Sandwich Panels, Part I: Design*

Eurocode 3 Annex A 1991, *Cold Formed Steel Sheeting and Members*

Eurocode 4 Part 1 1990, *Design of Composite Steel and Concrete Structures, Part 1—General Rules and Rules for Buildings*

Eurocode 4 Part 10 1990, *Design of Composite Steel and Concrete Structures, Part 10—Structural Fire Design*

Euronorm 148 1979, *Continuously Hot-Dip Zinc Coated Unalloyed Mild Steel Sheet and Coil with Specified Minimum Yield Strengths for Structural Purposes: Tolerances on Dimension and Shape*

Gough 1940, *The Stabilisation of a Thin Sheet by a Continuous Supporting Medium*

Harris 1969, *Local Failure of Plastic-Foam Core Sandwich Panels*

Hartsock 1969, *Design of Foam-Filled Structures*

Hartsock 1976, *Analysis of Sandwich Panels with Formed Faces*

Hassinen 1986, *Compression Strength of the Profiled Face in Sandwich Panels*

Hassinen 1990, *Design of Sandwich Panels against Thermal Loads*

Heagler 1987, *Composite Floor Deck in Tall Buildings*

Heagler 1989, *LRFD Design Manual for Composite Beams and Girders with Steel Decks*

Heagler 1991, *Design Practice for Steel Deck Systems*

Hoff 1956, *The Buckling of Sandwich-Type Panels*

Höglund 1986, *Load-Bearing Strength of Sandwich Panel Walls with Window Openings*

Karman 1940, *Mathematical Methods in Engineering*

Kozák 1991, *Steel-Concrete Structures for Multistory Buildings*

Kuenzi 1960, *Structural Sandwich Design Criteria*

Lawson 1990, *Good Practice in Composite Floor Construction*

prEN 10 147 1989, *Continuously Hot-Dip Zinc Coated Unalloyed Structural Steel Sheet and Strip; Technical Delivery Conditions*

SDI 1987a, *Design Manual for Composite Decks, Form Decks and Roof Decks*

SDI 1987b, *Steel Deck Institute Diaphragm Design Manual*

SDI 1987c, *Deck Damage and Penetrations*

SDI 1989, *Design Manual*

SIS 1991, *Composite Slabs, Practice Aspects*

Sokolnikoff 1956, *Mathematical Theory of Elasticity*

Tham 1982, *Flexural Bending and Axial Compression of Architectural Sandwich Panels*

Timoshenko 1951, *Theory of Elasticity*

Zienkiewicz 1967, *The Finite Element Method in Structural and Continuum Mechanics*

4

Shear Diaphragms

At the present time there exist a number of popular methods to calculate the capacity of a steel deck shear diaphragm. The most popular hand-calculation methods that have been proposed are (1) the Tri-Services method (TRI, 1973), (2) the Steel Deck Institute method (SDI, 1987), (3) the method by Bryan and Davies (1982), on which the European recommendations have been based (ECCS, 1977), and (4) the method by Easeley (1977). Not all of these methods, however, are suited to all types of diaphragms. The development of each method was primarily intended to cover a certain type of steel deck product and method of construction.

Even though a number of alternative design methods are available, in reality many of these methods are derivatives of the same basic philosophies. Except for the Tri-Services method, which is empirically based, the other three hand-calculation methods mentioned (SDI, Bryan and Davies, and Easeley) are almost identical in their basic formulation. This is particularly true of the strength calculations, but less so for the more complicated flexibility calculations.

Each method assumes the same basic failure modes or contributions to flexibility, and through a series of assumptions, simplifications, and nomenclature develops methods suited to a particular construction practice. For example, Bryan and Davies' design recommendations apply to roof decking or wall cladding profiles which are screw or blind-rivet fastened to each other, and where the connections to the supporting structure are screws or pins. A similar investigation into the SDI method shows that it is aimed at the thicker steel decking profiles where the side laps between decking sheets are actual laps usually screw-fastened together, and where the connections of the deck to the supporting structure are either welds, screws, or pins. In Canada, diaphragm bracing is almost exclusively done by the steel roof or floor deck and rarely by the wall cladding. The decking profiles used have a side lap which incorporates a male-female connection typically made by means of mechanical clinching (button punching), and arc spot welding is used to fasten the deck to the structural members.

A full examination and discussion of the similarities of the current diaphragm design methods is beyond the scope of this chapter. Section 4.1 describes the method incorporated in the provisions of SDI (1987); Section 4.2 is intended to propose an alternative design method more suited to current Canadian construction practice.

It is important to realize that the design expressions presented in Section 4.2 are a variation of the other design methods currently in use. It should also be noted that extensive laboratory testing has been carried out by each organization to substantiate their design expressions; similar tests to confirm the expressions presented in Section 4.2 have not been possible.

4.1 ROOF AND FLOOR DIAPHRAGMS (SDI METHOD)

The primary function of roof and floor systems is for resisting gravity loads and transferring their effects into the supporting structure. The choice of the structural elements used for resisting such loads is one of economics and depends on the type of framing systems used (see Section 3.1).

For steel-framed structures it is common to use cold-formed profiled steel decking for both roof and floor areas. Such decking is manufactured in a wide range of configurations and often allows multiple spans from a single decking length. Typical roof loads may dictate spans of perhaps 6 ft (1.8 m), using a relatively thin decking, whereas floors may require either shorter spans or thicker decking.

The more common profiled decking has depths between 38 and 76 mm (1.5 and 3.0 in.) with steel thicknesses of between 0.75 and 1.5 mm (0.0295 and 0.06 in.). Floor decking also may have embossments on some surfaces to enhance interlocking and composite slab behavior when concrete is used to finish floors. Roof areas may be finished using insulating concrete or some other method of overlay to form a smooth surface.

The individual deck must be properly attached to the structure since, during construction, they must sustain transient loads, including wind uplift forces. Decking-to-decking side-lap or "stitch" connections may be required to prevent edge separation between adjacent units. The latter are particularly important when concrete fill is to be used (see Section 3.1).

The complete assembly of decking, purlins, and framing members possesses beamlike characteristics and may be regarded in terms of a flexible deep girder. The assembly will act as a shear diaphragm and can be used to transfer in-plane forces between units of the structure.

The shear diaphragm bracing function often is considered a no-cost benefit, and this may be essentially true. The desired diaphragm strength and stiffness may be obtained with modest increases in the number of fasteners beyond those required for construction and gravity loading. The diaphragm assembly, however, serves as structural bracing and must be fabricated with the same attention to detail as any other bracing system.

1 Shear Elements

A horizontal diaphragm assembly involves the decking, any overlay or concrete fill, and structural members to which the decking is attached. Though its shear stiffness may be an order of magnitude lower than that of conventional girders, the diaphragm can be thought of as a thin-web girder. Each girder or diaphragm area is bounded by stiffeners, flanges on two sides, and shear walls on the other two sides. The web shear elements require that forces be transferred to them through the boundary members to which the shear elements must be properly attached.

The design issue usually is not one of developing all the shear potential for a particular assembly but rather to design connection details so that the assembly meets the bracing level required. The bracing level depends on perimeter conditions, the connections within the field, and diaphragm geometry.

2 Basic Diaphragm

It is clear that most diaphragms, when viewed in plan, will fall into the short, deep-beam category, as illustrated in Fig. 4.1. Further, the diaphragm shear stiffness often is an order of magnitude lower than that for a continuously connected flat plate of similar dimensions. Any analytical process, then, must include shear deflections, which may well dominate the total deflections for the system.

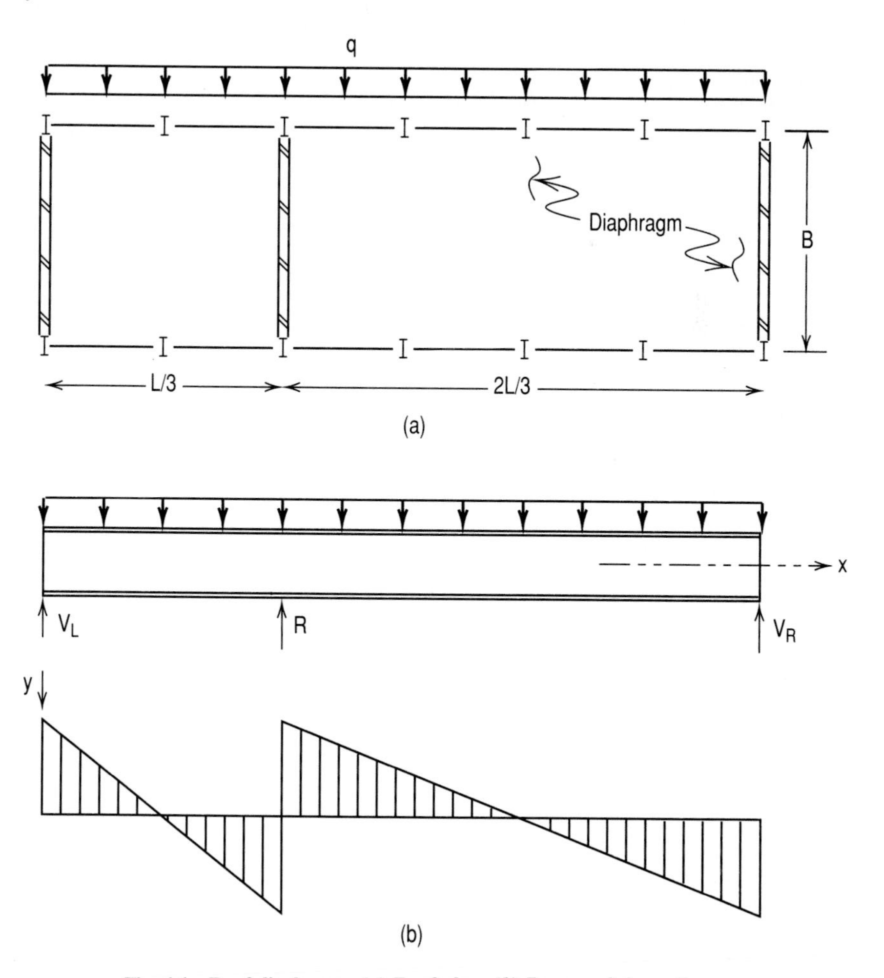

Fig. 4.1 Roof diaphragm. (*a*) Roof plan. (*b*) Beam and shear diagram.

Consider a roof area $B \times L$, as shown in Fig. 4.1. The diaphragm decks are not shown, but are to be supported on spandrel beams between columns on the long sides and on three shear walls in the shorter B direction. The system is to be evaluated for a uniformly distributed load q, that may have been either from a wind acting on the long walls or from an earthquake loading condition. Considering the entire roof area as a horizontal deep beam will require finding the shear wall reactions V_L, R, and V_R. From Fig. 4.2 these values may be found, noting that total deflection at the inner wall must be zero.

The change in shear deflection between two points may be found from changes in the shear diagram areas between those same two points. The shaded areas on the shear diagrams, divided by the shear width B and the stiffness G', establish the shear deflection at $x = L/3$. Further, presuming a beam moment of inertia I, the bending deflections can be found from a simple analysis.

Let the subscripts s and b indicate shear and bending deflections, respectively, while q and R represent uniform load and interior shear wall reaction. The following individual components of deflection result for $x = L/3$:

$$\Delta_{sq} = \frac{L}{3BG'}\left[\frac{1}{2}\left(\frac{qL}{2} + \frac{qL}{6}\right)\right] = \frac{qL^2}{9BG'} \tag{4.1}$$

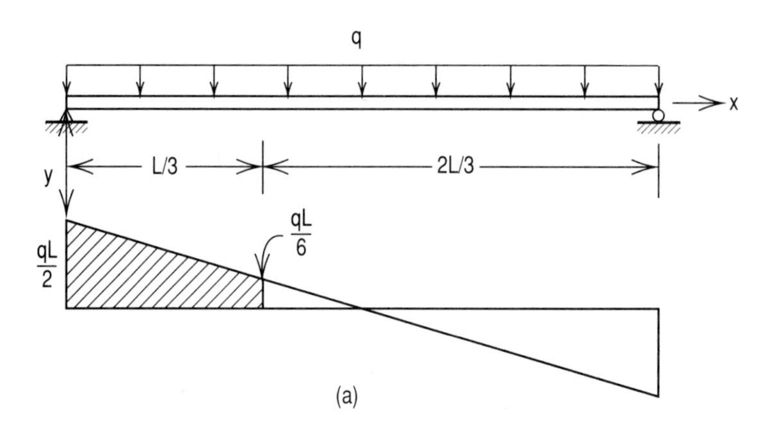

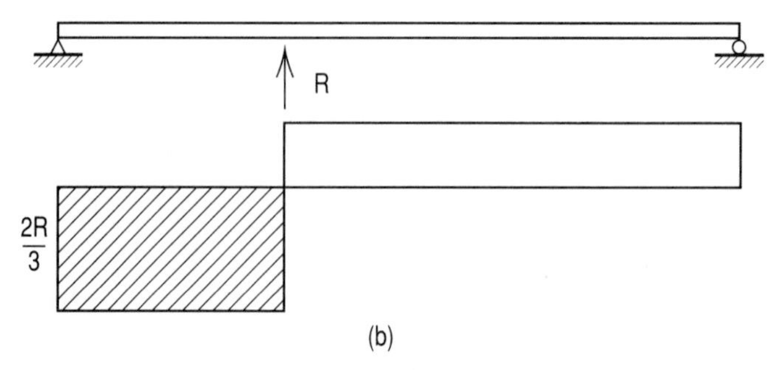

Fig. 4.2 Shear diagrams.

$$\Delta_{bq} = \frac{11qL^4}{972EI} \tag{4.2}$$

$$\Delta_{sR} = -\frac{1}{BG'}\left(\frac{2R}{3}\right)\frac{L}{3} = -\frac{2RL}{9BG'} \tag{4.3}$$

$$\Delta_{bR} = -\frac{4RL^3}{243EI} \tag{4.4}$$

where L = building length
B = building shear width
E = Young's modulus
q = uniform load
G' = diaphragm shear stiffness
R = interior wall reactions
I = beam moment of inertia

Considering the shear walls to be essentially rigid and defining $\alpha = EI/BG'$, the net deflection at the interior wall must be zero, resulting in

$$R = qL\left(\frac{108 + 11L^2/\alpha}{216 + 16L^2/\alpha}\right) \tag{4.5}$$

The interior wall reaction R is dependent on the relative stiffnesses α. If BG' is large relative to EI, the shear deflection component becomes small and R approaches $0.6875qL$, which is the reaction for a long slender beam on these spans. However, if G' is relatively small, R approaches $qL/2$ and the interior wall receives shear forces only from its tributary area, namely, half of each roof zone framing into it.

To underscore that the diaphragm deflection problem is largely one of shear only, consider a typical roof of thickness $t = 0.76$ mm (0.03 in.) and with a shear stiffness $G' = 7000$ N/mm (40 kips/in.). The equivalent shear modulus would be $G = G'/t = 9200$ N/mm^2 (1333 kips/in.^{2}) compared to a flat plate value of 78.3 kN/mm^2 (11,350 kips/in.^{2}). This diaphragm is on an order of magnitude more flexible in shear than is a flat plate.

The spandrel beams in Fig. 4.1 may not be so rigidly connected at their ends that they can act as continuous diaphragm flanges. However, their area A_s might be used to indicate an approximate lower-bound value for the girder moment of inertia, such as

$$I = 2A_s\left(\frac{B}{2}\right)^2 \tag{4.6}$$

As an example, consider the following data for the roof of a building as illustrated in Fig. 4.1: $B = 30$ m (98.5 ft), $L = 90$ m (295 ft), spandrel area $A_s = 9.68 \times 10^{-3}$ m^2 (15 in.^{2}), $E = 203 \times 10^6$ kPa (29,500 kip/in.^{2}), and $G' = 7 \times 10^6$ N/m (40 kips/in.). The "girder" moment of inertia may be approximated by:

$$I = 2A_s\left(\frac{B}{2}\right)^2 = \frac{A_s(B)^2}{2} \tag{4.7}$$

Then

$$\frac{EI}{BG'} = \frac{EA_sB}{2G'} = 203(9.68)(30)/(2 \times 7) = 4211 \text{ m}^2 \tag{4.8}$$

and

$$L^2/[EI/BG'] = 1.924$$

For the given relative stiffnesses and using Eq. 4.5, the interior wall reaction is:

$$R = qL\left(\frac{108+21}{216+31}\right) = 0.524qL \tag{4.9}$$

The interior shear wall reaction $R = 0.524qL$. In case of flange slip at the columns, the effective moment of inertia could be even smaller and the interior shear wall reaction is even closer to $0.5qL$. These values are typical for many diaphragms and indicate that they often can be treated as simple shear systems, ignoring bending deflections.

3 Diaphragm Field

The field of a diaphragm is independent of shear walls or braced frames where behavior is dictated by individual decks, their dimensions, and the type of connections used. In any decking of width w, connections may be made through some or all of the lower flat elements, as illustrated in Fig. 4.3. As external shear forces act in the plane of the diaphragm, they produce shear distortions as indicated in Fig. 4.4. Resistance to the shear distortion comes from a series of internal couples at cross-decking support members.

The Fig. 4.4 decks are shown with five structural connections engaging each deck at the end-of-decking purlins, four at each cross-decking interior purlins,

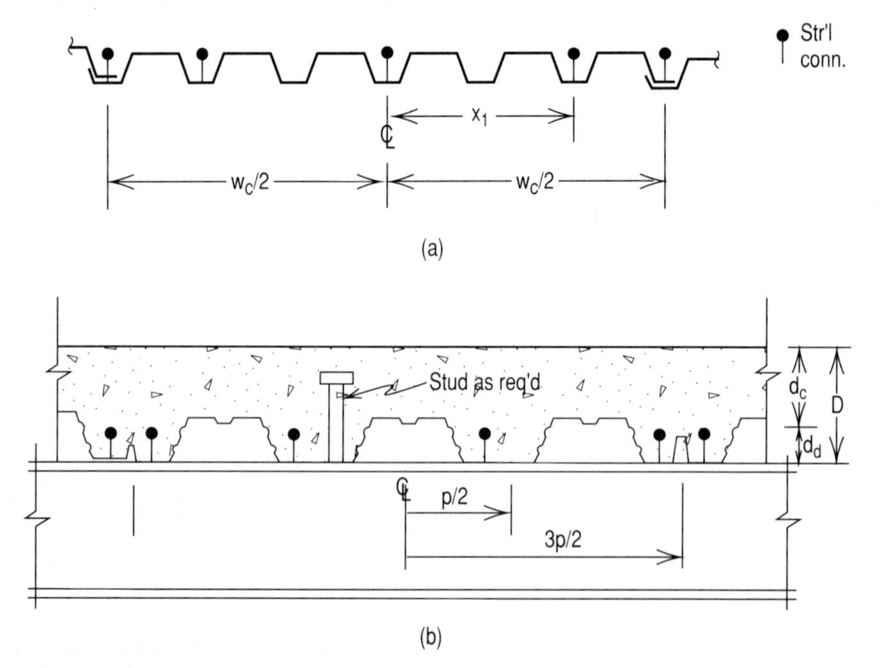

Fig. 4.3 Deck connections. (*a*) 36/5 fastener layout, $w_c = 914$ mm (36 in.). (*b*) Floor deck layout.

and two stitch connections per span in the side laps. Thus there is a total of six stitch connections per decking edge. The internal resisting couples depend on fastener position and fastener shear strength. Extensive research and development programs (SDI, 1987) have shown that the equilibrium of an interior decking near ultimate load can be expressed as

$$P_u \frac{w}{L_s} = 2M_e + n_p M_p + n_s Q_s w \tag{4.10}$$

where M_e = end of decking couple
M_p = interior purlin couple
L_s = decking length
n_p = number of interior purlins (2 shown)
n_s = number of stitch connections per decking (6 shown)
Q_s = stitch connector shear strength

With a linear variation in the connector shear forces F_p across the decking and

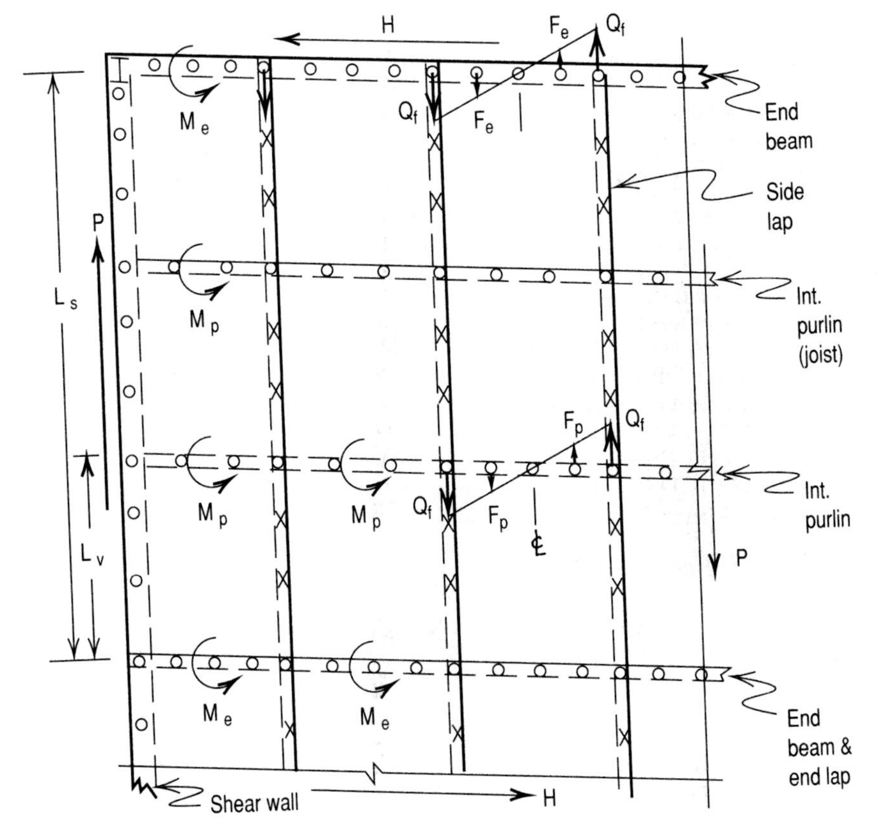

Fig. 4.4 Layout indicating internal resisting couples.

relative to the limiting fastener shear strength Q_f,

$$F_p = Q_f\left(\frac{X_p}{w/2}\right) \tag{4.11}$$

with

$$M_p = \sum F_p X_p = \frac{2}{w} Q_f \sum x_p^2 \tag{4.12}$$

$$M_e = \frac{2}{w} Q_f \sum x_e^2 \tag{4.13}$$

The end-of-decking couple may be somewhat limited depending on the ability of the edgemost corrugation to resist eccentric compression. It has been found (SDI, 1987; Luttrell, 1981) that the one connection at the compression corner is limited through

$$\theta = 1 - \frac{DL_v}{C\sqrt{t}} \tag{4.14}$$

where D = decking depth, mm (in.)
L_v = purlin spacing, m (ft)
t = decking thickness, mm (in.)
C = 370 (240 in U.S. units)

Equation 4.12 then can be reduced to the average ultimate shear strength per unit length,

$$S_u = \frac{P_u}{L_s} = [2(\theta - 1) + B]\frac{Q_f}{L_s} \tag{4.15}$$

where

$$B = n_s\alpha_s + \frac{1}{w^2}\left(2n_p \sum x_p^2 + 4\sum x_e^2\right) \tag{4.16}$$

$$\alpha_s = \frac{Q_s}{Q_f} \tag{4.17}$$

When the steel deck is finished with structural concrete fill, the θ term becomes unity, indicating that corner buckling is eliminated because of alignment forced by the concrete. Further, an additional path of shear transfer through the concrete is present. The concrete forms a bond with the decking and retards decking warping and side-lap slip. This additional path in the diaphragm field can greatly increase shear strength and can lead to a system similar in shear strength to that of a conventional flat slab. With a minimum concrete depth of cover, d_c = 63 mm (2.5 in.), Eq. 4.15 then is modified to

$$S_u = \frac{BQ_f}{L_s} + \frac{w^{1.5}/\sqrt{f_c}}{C_d} \quad \text{(kips/ft)} \tag{4.18}$$

where w = concrete unit weight, kg/m^3 (lb/ft^3)
f_c = concrete compressive strength kPa (lb/in.^{2})
C_d = 225,000 (19,500 in U.S. units)

Lightweight insulating concrete and other finishing overlays also provide additional shear paths, but at considerably lower levels than indicated for normal concrete by Eq. 4.18. SDI (1987) contains formulations for such systems.

4 Diaphragm Perimeter

Figure 4.4 indicates two perimeter conditions. One is at the decking ends and the other is over a shear wall. The decking end fastener arrangement has already been given consideration as part of the field evaluation in Eq. 4.16.

The decking along the shear wall can be critical to shear transfer, particularly if the diaphragm is concrete-filled and the concrete must be terminated over at the wall. Concrete termination would remove a shear path that was present in the diaphragm field. The first decking must transfer its shears through couples at the purlins and through the side-lap connections along the first interior side lap. On the other side of the first decking and over the wall, it is essential that decking-to-wall connections develop the needed capacity. Figure 4.4 indicates edge fasteners with the same spacing as those along the first interior side lap. Connections from a decking to the supporting framework usually are stronger than side-lap connections, and spacing them to match side-lap spacing would be conservative for nonfilled diaphragms. Heavily loaded composite floor diaphragms may require shear studs on the perimeter.

At the ends of the decking, fasteners may be much more closely spaced than along diaphragm edges over shear walls. These spacing effects are considered within the formulations presented, except when different types of connectors are used. The design issue is simply to determine the ultimate average shear force along the top of the wall and to space uniformly an adequate number of connectors of known strength. Within the first decking, all other fastener arrangements should match those in the second decking.

Two special edge conditions arise. When the diaphragm is supported on joists that rest on top of the shear wall, there is no ready method for installing edge connections *between* purlins. It is common practice to extend a thin steel angle along the shear wall and weld it to each joist top. The angle can then support decking edges and provide locations for frequent edge connections. The shear transfer through the joist end may require increased welding at the wall top.

A second edge condition arises when concrete-filled diaphragms are used. The first interior side lap, and other components in the field, are strengthened by the concrete shear path. The concrete will terminate at the shear wall, leaving this shear line weaker than others. Some edge enhancement may be developed by spacing the edge connections more closely, but if very high shears are present, shear studs may be required to transfer concrete shear forces directly into the shear walls.

5 Connector Shear Strength

The most common method of connecting diaphragm decking to the structure is through arc "puddle" welding. An electrode is used to blow a hole in the decking and welding is continued by "puddling" the weld material in the hole. Such operations require intimate contact between connected parts to facilitate heat transfer, and the operation requires considerable skill. Such a welding process typically is limited to decking of 22 gage [0.75 mm (0.0295 in.)] or heavier.

Structural welds through thin decking can be accomplished using welding washers which have prepunched holes. Welds are made through the hole. As welding begins, the washer absorbs heat and limits burnout in the decking. Once the washer is "puddled" full of weld material, the operation stops and the weld cools and clamps the washer onto the decking. Any welding operation demands that the welding continue until the supporting structural member has reached fusion temperature, often for several seconds.

Values for diaphragm shear-loaded welds are shown in several references (SDI, 1987; Luttrell, 1981; AWS, 1981; AISI, 1986) and take the general form

$$\text{Puddle welds:} \qquad Q_f = 2.2 \times 10^{-3} t F_u (d - t) \quad \text{(kN)} \tag{4.19}$$

$$\text{Weld washers:} \qquad Q_f = 680 \times 10^{-6} t (1.33 d_0 + 0.044 F_{xx} t) \quad \text{(kN)} \tag{4.20}$$

where Q_f = weld strength, kN
d = visible weld diameter, mm
d_0 = washer hole diameter, mm
F_u = steel strength, kPa
F_{xx} = electrode strength, kPa
t = decking thickness, mm

Several types of powder-actuated fasteners and pneumatically driven pins have been developed for decking-to-structure connections. These have the advantage of consistent quality, adaptability to use in cold or wet environments, and more rapid installation. Shear values for such connections can be supplied by the manufacturer.

Decking side-lap or stitch connections most often are made using self-drilling screws. They are preferred over welding because sheet-to-sheet welding is difficult and of questionable quality. Some decks are manufactured with an upstanding and overlapping seam suitable for button punching. In this operation, a crimping tool is used to form a series of nested conelike indentations. On removing the tool, elastic rebound will allow some loosening in the cones. Such connections may be adequate to prevent vertical separation at the sheet edges but will be much weaker in shear than screws.

Sheet-to-sheet screws under shear load usually are limited in strength by tearing or splitting in the sheet around the screw. The screw strength (SDI, 1987) shows only moderate sensitivity to the material strength and may be represented by

$$Q_s = 0.8(dt) \quad \text{(kN)} \tag{4.21}$$

where d = screw diameter, mm
t = diaphragm thickness, mm

For a typical No. 12 screw, $Q_s = 24.3t$ kips. Screw manufacturers can supply the designer with specific screw data. In Europe a different formula has been used, as shown in Chapter 5.

6 Load Factors

Detailed studies have been made (Luttrell, 1981) based on the results from many full-scale diaphragm tests undertaken over the last 30 years and using various connection means. These have led SDI (1987) to relate working shears S from transient loads to the foregoing formulations for ultimate strength as follows:

$$S = \frac{S_u}{SF} \tag{4.22}$$

where SF = safety factor
$SF = 2.75$ for welded connections
$SF = 2.35$ for mechanical connections
$SF = 2.75$ for weld-screw combinations
$SF = 3.25$ for concrete-filled systems

7 Composite Floor Diaphragms

Floor diaphragms in high-rise buildings may be required to transfer lateral loads from the floor to short shear-resistant zones such as those adjacent to central service shafts in a building. Consider the diaphragm arrangement shown in Fig. 4.5*a*, which represents the upper floors of a building. The columns may be considered flexible in shear relative to the core, and lateral support then is provided through the two vertical shafts indicated. Any floor level can be viewed in plan, as in Fig. 4.5*b*. The lateral line load q represents either wind pressures on walls connected to the floor or earthquake effects from the mass at the particular level.

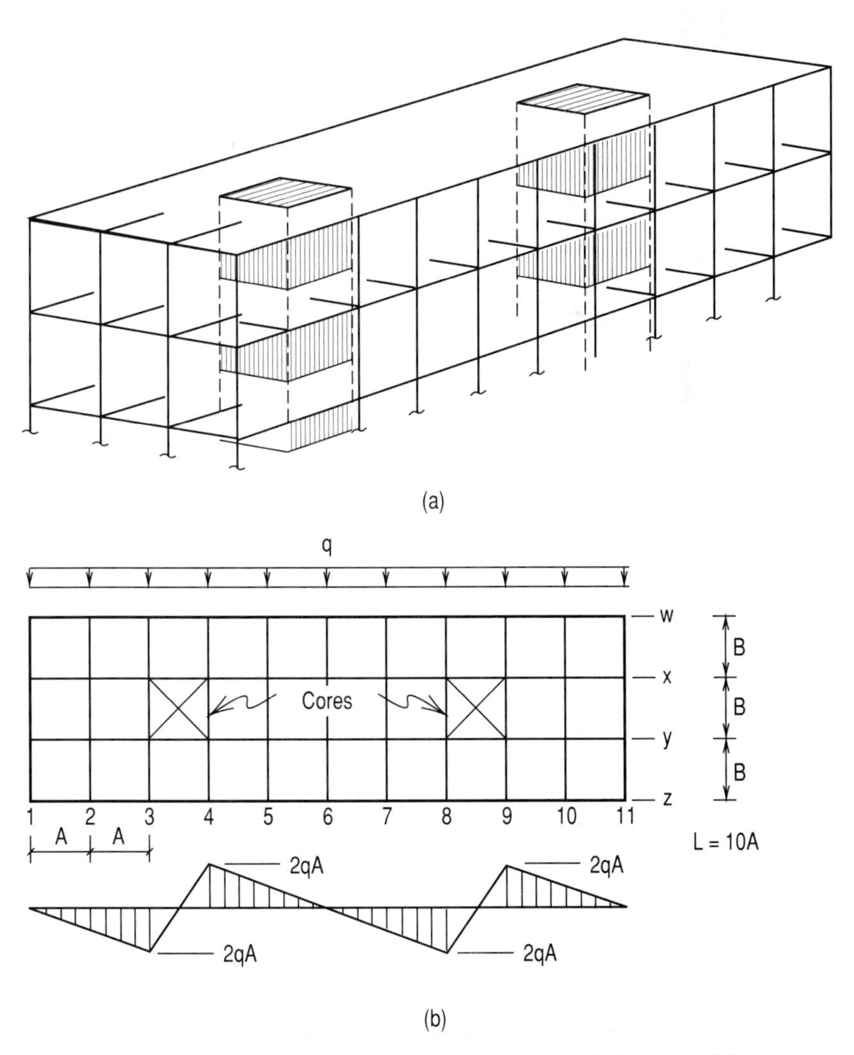

Fig. 4.5 (*a*) Floor diaphragms with dual shear cores. (*b*) Plan view of floor.

The line load q produces a shear diagram, as shown in Fig. 4.5b, where the maximum shear ordinate is $q(2A)$. Between column lines 2 and 3 or between lines 9 and 10 and near the cores, the average maximum diaphragm shear is $2qA/3B$. If it is presumed that the core areas develop most of their resistance in vertical planes parallel to the B dimensions, these planes will be resisting forces of $5qA/2$. This leads to average diaphragm shears of $5qA/2B$ in the short transfer zone at the core, and these are slightly greater than three times the average maximum values between lines 2 and 3.

Consider the floor plan in Fig. 4.5b with a width of $3B = 33$ m (108 ft) and with $L = 10A = 90$ m (295 ft). The story height is 4 m (13 ft). A wind study has identified a windward design pressure of 750 Pa (15.7 psf) and a leeward pressure of 480 Pa (10 psf). The net down-wind diaphragm line load is $q = (750 + 480)(4) = 4.92$ kN/m (0.337 kips/ft). The maximum average shear to the left of line 3 is $2q\Lambda/3B = 2(4.92)(9)/(33) = 2.684$ kN/m (0.184 kips/ft).

Adjacent to the core in line 3, the average shear is $5q\Lambda/2B = 10.06$ kN/m (0.689 kips/ft). With a load factor set at 3.25, this evaluation would require the diaphragm designs:

$$\text{Field:} \quad S_u = 3.25(2.684) = 8.72 \text{ kN/m } (0.598 \text{ kips/ft})$$

$$\text{Core edge:} \quad S_u = 3.25(10.06) = 32.70 \text{ kN/m } (2.240 \text{ kips/ft})$$

A preliminary design has resulted in the following selections:

Steel

Decking depth,	$D_d = 50$ mm (2 in.)
Decking width,	$w_c = 912$ mm (36 in.)
Corrugation width,	$p = 304$ mm (12 in.)
Decking thickness,	$t = 0.9$ mm (0.035 in.)
Decking length,	$L_s = 8.25$ m (27 ft.)
Six spans in L_s,	$L_v = 1.375$ m (4.5 ft.)
No. int. purlins,	$n_p = 5$
Steel strength,	$F_u = 310 \times 10^3$ kN/m^3 (45 kips/in^2)

Concrete

Concrete strength,	$f_c = 20{,}700$ kPa (3000 lbs/in^3)
Concrete weight,	$w = 2400$ kg/m^3 (150 lbs/ft^3)

Connections

16 mm ($^5/_8$ in.) diameter arc puddle welds at all supports.

All sidelap stitch connections are No. 12 screws.

$$\begin{aligned}\text{Supports: } Q &= 2.2 \times 10^{-3} F_u(t)(d - t) \\ &= 2.2 \times 10^{-3}(310)(0.9)(16 - 0.9) \\ &= 9.268 \text{ kN } (2.084 \text{ kips})\end{aligned} \tag{4.23}$$

No. 12 Side laps screws have a major diameter of 5.35 mm and, following Eq. 4.21, the screw strength is

$$\begin{aligned}Q &= 0.8\,dt = 4.28(0.9) \\ &= 3.85 \text{ kN } (0.865 \text{ kips})\end{aligned} \tag{4.24}$$

$$\alpha = Q_s/Q_f = 0.42 \tag{4.25}$$

At all cross supports, welds are located near the center of each lower decking element at an approxmate spacing p. Then for Eqs. 4.11, 4.13, and 4.15,

$$\sum (x_l/w_c)^2 = \sum (x_p/w_c)^2 = 5p^2/(3p)^2 = 0.556 \tag{4.26}$$

Eq. 4.16: $B = 6(2)(0.42) + [2(5) + 4](0.556) = 12.824$

Eq. 4.14: $\Theta = 1 - 50(1.375)/(370\sqrt{t}) = 0.814$

Prior to placing the concrete, the diaphragm strength is found from Eq. 4.15:

$$S_u = [2(\Theta - 1) + B](Q_f/L_s) = 13.99 \text{ kN/m } (0.957 \text{ kips/ft}) \tag{4.27}$$

After the concrete is cured, the diaphragm field strength is found from Eq. 4.18:

$$\begin{aligned} S_u &= 12.824(9.268/8.25) + [(2400)^{1.5}(20700)^{0.5}]/225000 \\ &= 89.6 \text{ kN/m } (6.14 \text{ kips/ft.}) \end{aligned} \tag{4.28}$$

If the diaphragm longitudinal edges are welded to match the interior sidelap spacing, these welds are spaced at 450 mm (18 in.) leading to 2.22 welds per meter including those at the cross-support ends. With no other edge attachments and whether or not the diaphragm is filled with concrete, the maximum edge shear strength is,

$$S_u = 2.22Q_f = 2.22(9.268) = 20.57 \text{ kN/m } (1.41 \text{ kips/ft.}) \tag{4.29}$$

The design requirement for the field was 8.72 kN/m and, at the core transfer line, the requirement was 32.70 kN/m. Obviously the diaphragm is satisfactory in all areas except adjacent to the core where it will be necessary to increase the number of edge connectors. Alternately, it may be better to make beam lines, adjacent to the core, axially continuous to allow for a more uniform shear transfer over a longer distance. Along those frame lines adjacent to the cores and in lines X and Y, it is essential that beams be properly detailed to avoid relative end movements. Though the axial forces in these members from collected diaphragm shears may be small, any slip at the core corners can cause high local stresses in the diaphragm and possible connection failures.

8 Commentary

The design of steel deck diaphragms is not particularly difficult when the system is kept in proper perspective. The diaphragm simply is a large girderlike system subject to all the limitations of any other shear-resistant system. Care must be taken in its design, with particular attention given to connection details and load transfer lines. It is a bracing system to be designed and erected with the same care as any other structural system.

4.2 DESIGN METHOD FOR STEEL DECK SHEAR DIAPHRAGMS (PROPOSED METHOD FOR CANADIAN PRACTICE)

1 Basic Assumptions

Failure Modes. The formulas developed here assume a diaphragm assembly as shown in Fig. 4.6. This arrangement is typical of most theories of diaphragm strength. As a load is applied to one corner of the diaphragm, a racking force is developed in the structure. This racking force is resisted by the steel decking acting as a diaphragm.

If the deck were actually a flat plate, a structure loaded like this would develop a diagonal tension field in the plate, eventually resulting in a shear buckling failure (assuming the fasteners did not fail first). A shear buckling failure mode can be experienced in diaphragms that use very thin, lightly profiled cladding sheets. A design provision for this type of failure mode is given by Bryan and Davies (1982), who investigated cladding profiles in addition to decking profiles. Steel deck, however, is usually subjected to a significant dead load acting perpendicular to the diaphragm shear forces, which will tend to restrain shear buckling. In addition, the relatively thick sheet steel and the deeply profiled shapes will reduce the possibility of shear buckling as a potential failure mode. Consequently, the failure modes that will be used for the prediction of the diaphragm strength are strictly fastener failures.

As a shear force is applied to the diaphragm, the fasteners must transfer this force from the point of load application into the diaphragm sheet and then back into the structure at the braced points. In theory there are many possible failure modes, but in practice there are three dominant ones, which will be addressed: (1) failure of the fasteners along the edge of the diaphragm, (2) failure of the fasteners along a seam between deck sheets, and (3) failure of the fasteners at the corners of the diaphragm. The ultimate strength of the diaphragm will be the lowest calculated capacity of the three failure modes considered.

Fastener Failure. Throughout this analysis it will be assumed that the fasteners exhibit a linearly elastic load deformation behavior up to their failure load. If this

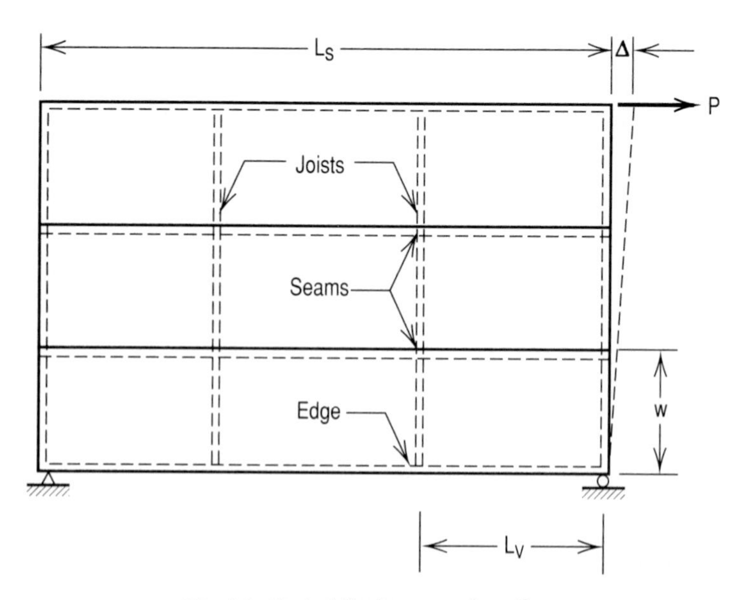

Fig. 4.6 Typical diaphragm configuration.

is the case, the following equation would apply to all fasteners:

$$F = K\Delta \tag{4.30}$$

where F = shear force developed in fastener
K = fastener stiffness
Δ = deformation of fastener under load

If only one type of fastener is used throughout the diaphragm, the design equations can be simplified. Often, however, different fasteners are used to make different connections. For example, the connections along the seams may be done by mechanical clinching, and the fastening of the deck to the supporting structure may be done by arc spot welding. The equations that will be developed to predict the diaphragm capacity assume that the failure of the diaphragm occurs when the first fastener fails. When determining the failure load of the diaphragm, it is necessary to select a specific fastener upon which to base the analysis.

One mode of failure for a diaphragm is the failure of a seam in shear. This failure mode requires the seam to have deformed sufficiently to transfer enough load into the seam fasteners to cause one of them to fail. The equation for the seam strength could be expressed in terms of a button punch connection as follows:

$$p = n_b F_b + \frac{K_w F_w n_w}{K_b} \tag{4.31}$$

where K = fastener stiffness
n = number of fasteners
F = load in fastener

and the subscripts b and w denote button punch and arc spot weld respectively.

In order for a button punch along the seam to fail, it must deform an amount Δ_s, this deformation being much greater than the deformation Δ_w required to fail the weld. Therefore, before the button punch can be deformed sufficiently to cause it to fail as assumed in Eq. 4.31, the weld will fail at a lower displacement and, consequently, at a lower p.

The conclusion that can be drawn from this example is that the load-displacement curves of the various fasteners need to be considered when deriving the equation for the strength of a diaphragm. The strength of the diaphragm will be contingent on the fasteners that will fail first under the resulting deformation. In reality, a fastener exhibits a more elastic-plastic load deformation behavior, which allows for more load sharing among fasteners. However, this mechanism is not addressed by the present design methods.

2 Strength Calculations

In this section the following symbols will be used:

F_e = shear strength of end fasteners (in most cases this fastener is a weld through two sheets of deck onto the supporting structure)
F_s = shear in seam fastener = $F_e K_s / K_e$
F_{st} = shear in shear transfer fastener = $F_e K_{st} / K_e$
F_{ue} = ultimate strength of end fastener (weld through two sheets)
F_{uw} = ultimate strength of interior fastener (weld through one sheet)

F_w = shear in interior fastener (weld through one sheet)
K_e = stiffness of end fastener
K_s = stiffness of seam fastener
K_{st} = stiffness of shear transfer fastener
K_w = stiffness of interior fastener
L_s = length of deck sheet
L_v = length of deck span (joist spacing)
n_p = number of structural supports $= L_s/L_v + 1$
n_s = number of seam fasteners along one seam of deck sheet $= L_s/s$
n_{st} = number of shear transfer connections along entire edge of deck sheet
s = spacing of seam fasteners
V = applied shear per unit length
w = width of deck sheet
x_i = distance from centerline of deck to transverse fasteners (end or interior fasteners)
z = number of welds across deck width
α = number of shear transfer elements between joists
β = number of continuous spans $= L_s/L_v$

Fastener Failure in Edge Members. The first failure mode considered is the possibility of fastener failure along the decking edges. Refer to the diaphragm illustrated in Fig. 4.6 and consider it as just a part of a larger diaphragm. The edges of the diaphragm must be connected adequately to transfer the applied load into the diaphragm along the one edge, and then out of the diaphragm and into the supporting structure along the other edge. In a real diaphragm there will be many individual deck sheets along the diaphragm edge which are lapped at their ends. A weld through two deck sheets at this end lap will have a different capacity than a weld through only one sheet. This difference is reflected in the analysis that follows. The decking forces along the diaphragm edge are illustrated in Fig. 4.7. The equilibrium of these forces is

$$P = 2F_e\left(\frac{2}{w}\right)\left(\sum_{i=1}^{z} x_i\right)\left(\frac{1}{2}\right) + (n_p - 2)F_w\left(\frac{2}{w}\right)\left(\sum_{i=1}^{z} x_i\right)\left(\frac{1}{2}\right) + n_{st}F_{st} \tag{4.32}$$

Since the load on a fastener is assumed to be proportional to its stiffness, and also assuming that the stiffness of a weld through two sheets is the same as through one sheet, the load carried by a single-sheet weld will equal the load carried by a double-sheet weld. To simplify, therefore, set $F_w = F_e$, since the limiting load will be determined by failure in the weld through two sheets at the end lap.

Assuming that the weld pattern is maintained along the entire length of the sheet, Eq. 4.32 simplifies to

$$P = n_pF_e\left(\frac{2}{w}\right)\left(\sum_{i=1}^{z} x_i\right)\left(\frac{1}{2}\right) + \beta\alpha F_{st} \tag{4.33}$$

If it is assumed that the edge of the deck is welded to a structural member, then the number of shear transfer elements becomes the number of welds along this edge between the joists (not including the welds at the joists). Equation 4.33 then simplifies to

$$P = \frac{n_pF_e}{w}\left(\sum_{i=1}^{z} x_i\right) + \beta\alpha F_e \tag{4.34}$$

Substituting into Eq. 4.34 $V = P/L_s$ (the shear per unit length) gives

$$V = \frac{F_e}{L_s}\left[\frac{n_p}{w}\left(\sum_{i=1}^{z} x_i\right) + \beta\alpha\right] \tag{4.35}$$

The limiting capacity of the diaphragm determined by this failure mode is calculated by setting F_e equal to its ultimate capacity F_{ue}.

Fastener Failure in Interior Seam. Figure 4.8 shows the distribution of fastener forces for a single deck sheet within the diaphragm. By taking the equilibrium

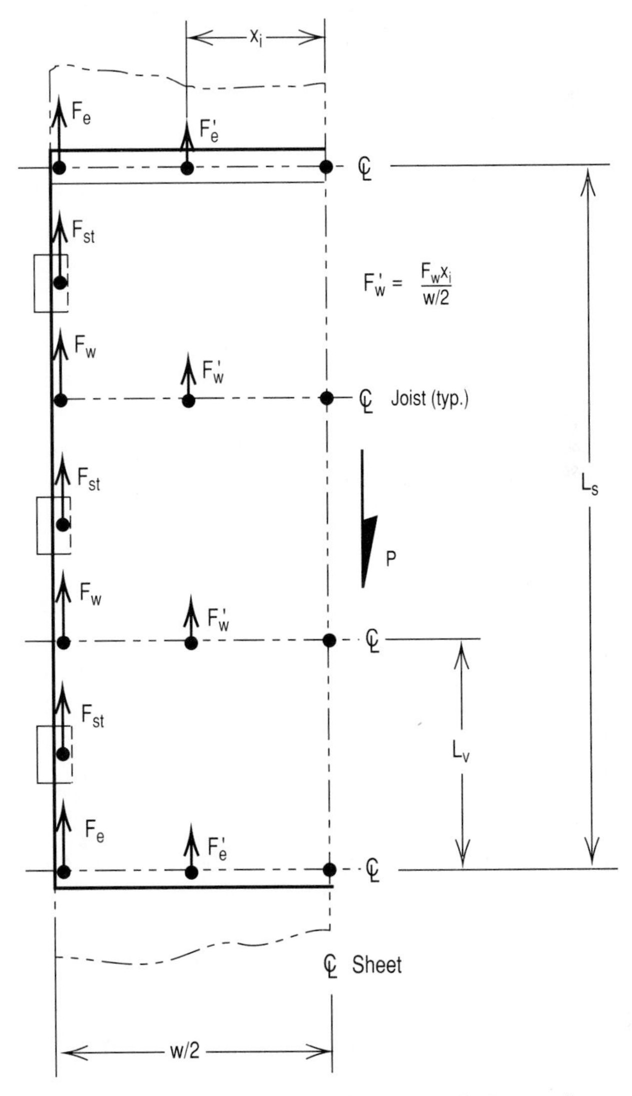

Fig. 4.7 Distribution of fastener forces along diaphragm edge.

about one corner of the deck sheet, these fastener forces can be translated into a force-moment distribution, which can be written as follows:

$$L_s V w = w F_s n_s + 2M_e + (n_p - 2)M_w \tag{4.36}$$

where

$$M_e = \frac{2}{w}\left(\sum_{i=1}^{z} x_i^2\right)F_e \tag{4.37}$$

$$M_w = \frac{2}{w}\left(\sum_{i=1}^{z} x_i^2\right)F_w \tag{4.38}$$

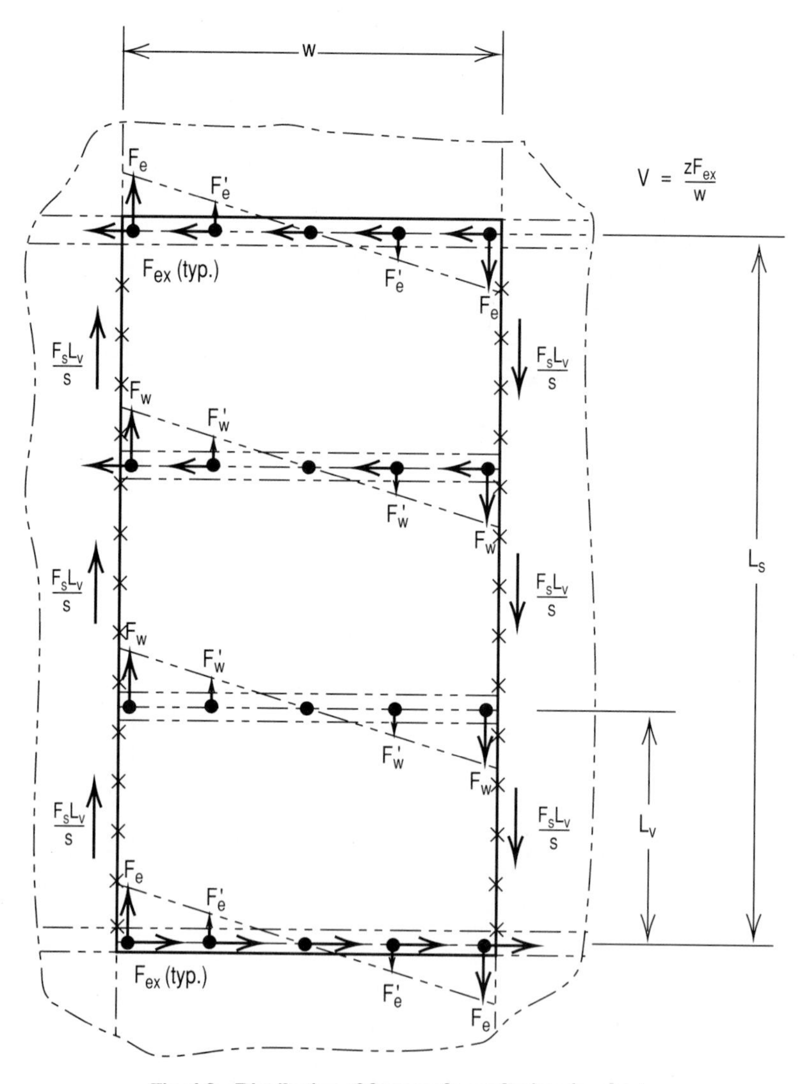

Fig. 4.8 Distribution of fastener forces for interior sheet.

Substituting the expressions for M_e and M_w into Eq. 4.36 gives

$$V = \frac{1}{L_s w}\left[w n_s F_s + 2F_e \frac{2}{w}\left(\sum_{i=1}^{z} x_i^2\right) + (n_p - 2)F_w \frac{2}{w}\left(\sum_{i=1}^{z} x_i^2\right)\right] \tag{4.39}$$

If the assumption is made that $F_w = F_e$ for reasons stated previously, then Eq. 4.39 reduces to

$$V = F_e\left[\frac{K_s}{sK_e} + \frac{2}{w^2}\frac{n_p}{L_s}\left(\sum_{i=1}^{z} x_i^2\right)\right] \tag{4.40}$$

The ultimate strength of the diaphragm V_u for this failure mode would be calculated by setting F_e equal to the ultimate strength of an arc spot weld made through two sheets, F_{ue}.

Fastener Failure at Diaphragm Corners. A steel deck acting as a diaphragm must transfer the applied forces from the diaphragm into the structural framing. A diaphragm considered in its entirety will have uniform shear along all four sides. The transfer of shear occurring in the fasteners will be distributed to each fastener according to its relative stiffness. However, at the diaphragm corners the corner fastener is acted upon by the shear forces along both edges. The resultant of the two forces acting on this fastener, as illustrated in Fig. 4.9, will be a limiting capacity for the diaphragm.

The magnitude of force F_1 is taken as the average shear on the fasteners along the edge of the diaphragm,

$$F_1 = V\left(\frac{w}{z}\right) \tag{4.41}$$

The force on the corner fastener F_2 is computed in the same manner as in the previous section, where the forces in the seam fasteners were calculated. F_2 is determined from Eq. 4.40,

$$F_2 = \frac{V}{C_1} \tag{4.42}$$

where

$$C_1 = \left[\frac{K_s}{sK_e} + \frac{2}{w^2}\frac{n_p}{L_s}\left(\sum_{i=1}^{z} x_i^2\right)\right] \tag{4.43}$$

The resultant of forces F_1 and F_2 equals

$$F_r = \sqrt{F_1^2 + F_2^2} \tag{4.44}$$

$$F_r = V\left[\frac{1}{(z/w)^2} + \frac{1}{C_1^2}\right]^{1/2} \tag{4.45}$$

Since the deck sheets are not lapped at the corners, the capacity of the diaphragm for this failure mode is determined by setting F_r equal to the ultimate strength of a weld through one sheet, F_{uw}. For this failure mode the equation for the ultimate strength of the diaphragm then equals

$$V_u = \frac{F_{uw}}{[1/(z/w)^2 + 1/C_1^2]^{1/2}} \tag{4.46}$$

3 *Flexibility Calculations*

The work of Bryan and Davies (1982) and SDI (1987) have proposed very comprehensive treatments of the diaphragm flexibility. These methods are almost identical in their analysis, but appear superficially different, as was the case in the development of the strength calculations. There is, however, one fairly significant difference in the determination of the flexibility due to the warping of the profile.

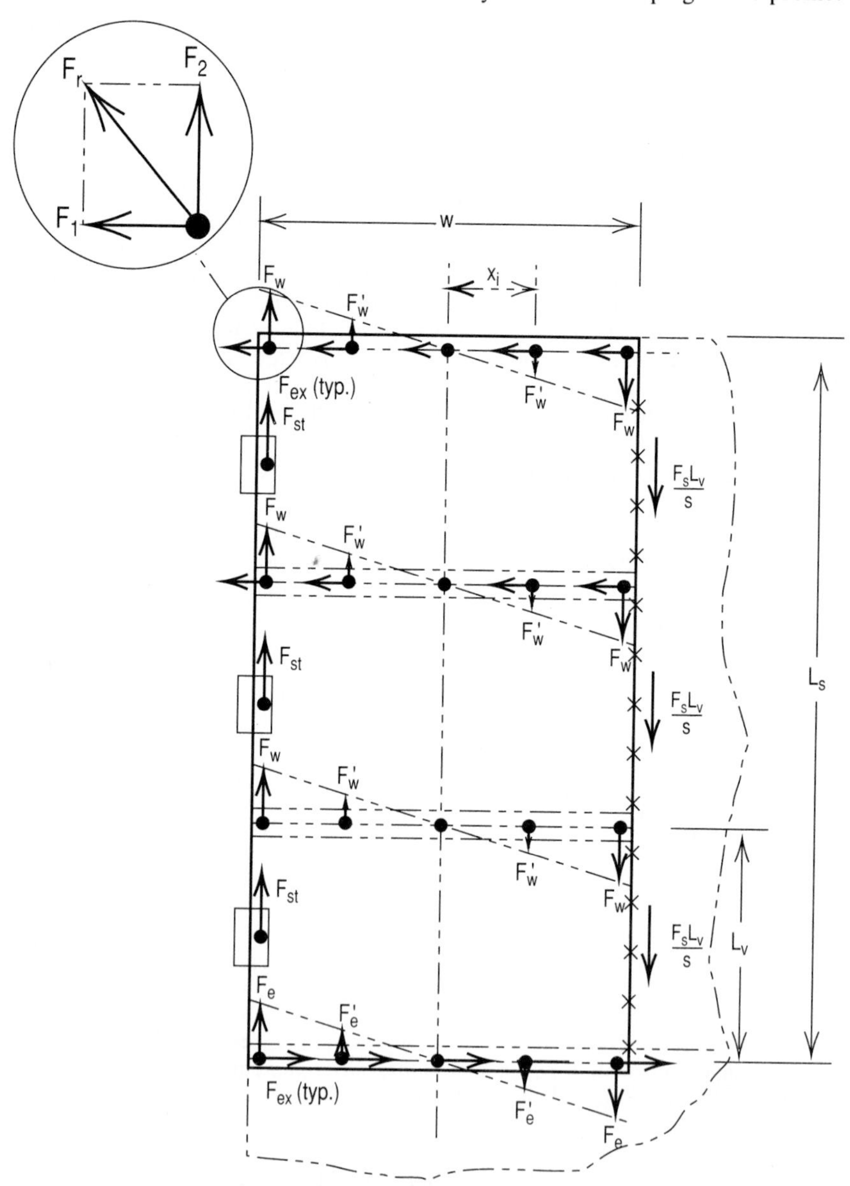

Fig. 4.9 Distribution of fastener forces for corner sheet.

This is a very complex problem to calculate by hand and is better solved by a finite-element analysis. Bryan and Davies have done such an analysis and provided tabulated values for a warping constant which is specific to the deck profile geometry. For this reason, the Bryan and Davies method for flexibility calculations has been adopted in principle. The determination of the diaphragm flexibility is as follows.

The flexibility of a diaphragm is obtained by considering each of the component flexibilities in turn and then summing them to give the total flexibility of the diaphragm. The total flexibility is the sum of the following components:

1. Sheet deformation
   - $c_{1.1}$ = flexibility due to distortion of deck profile
   - $c_{1.2}$ = flexibility due to shear strain in sheet
2. Fastener deformation
   - $c_{2.1}$ = flexibility due to fastener slip along seams
   - $c_{2.2}$ = flexibility due to fastener slip along diaphragm edges
   - $c_{2.3}$ = flexibility due to fastener slip perpendicular to deck span
3. Flexural deformation
   - c_3 = flexibility due to flexural action of equivalent plate girder

The specific equations for each of these flexibility components will not be reproduced here. Reference should be made to Bryan and Davies (1982), which presents the procedure. The equations developed by Bryan and Davies will need to be modified, however, to accommodate the nomenclature and diaphragm construction details that have been assumed earlier.

4.3 CONCLUSIONS

Possibly the most interesting conclusion from the investigation into the design of steel deck diaphragms is the similarity among the current design methods. Current understanding among most of the engineering design profession is that there are many methods available, all of which give different estimates of strength and stiffness. While it is true that there are different methods available, the most popular hand-calculation methods are very similar. In retrospect, this conclusion should not be that surprising. The failure modes observed during various independent testing reveal a consistency which has led to the derivation of the design methods being proposed. Given that there are only a few dominant failure modes, the different design methods have to address these limit states in a like manner.

Section 4.2 is a summary of an alternative method for designing steel deck shear diaphragms. The intention is to provide the structural designer with a method for designing diaphragms which is easily understood. Except for the few designers and researchers who are familiar with this topic, the design of steel deck shear diaphragms remains a mystery to many engineers.

4.4 CONDENSED REFERENCES/BIBLIOGRAPHY

AISC 1989, *Manual of Steel Construction*

AISI 1986, *Specification for the Design of Cold-Formed Steel Structural Members*

AWS 1981, *Structural Welding Code, Sheet Steel*

Bryan 1982, *Manual of Stressed Skin Diaphragm Design*

Easeley 1977, *Strength and Stiffness of Corrugated Metal Shear Diaphragms*
ECCS 1977, *European Recommendations for the Stressed Skin Design of Steel Structures*

Fox 1986, *A Design Method for Steel Deck Shear Diaphragms*

Luttrell 1980, *Theoretical and Physical Evaluations of Steel Shear Diaphragms*
Luttrell 1981, *Steel Deck Institute Diaphragm Design Manual*
Luttrell 1991, *Roof and Floor Diaphragms*

SDI 1987, *Steel Deck Institute Diaphragm Design Manual*

TRI 1973, *Seismic Design for Buildings*
TRI 1982, *Seismic Design for Buildings*

5

Connections in Cold-Formed Steel

Connections are an important part of every structure, not only from the point of view of structural behavior, but also in relation to the cost of production (Council on Tall Buildings, Group SB, 1979). It has been shown that for a structure of hot-rolled sections, about 40% of the total cost is directly or indirectly influenced by the connections. There is no reason to believe that the percentage will be much lower for thin-walled structures. For economic reasons the influence of the joining process on the cost will tend to increase (Tomà and Stark, 1982).

A variety of joining methods are available for thin-walled structures. In the first three sections this chapter surveys the most frequently used mechanical fasteners and provides information on welding and adhesive bonding. Section 5.4 treats the structural and nonstructural requirements that connections have to fulfill. Depending on these requirements, a procedure is given to select the fastening system. Section 5.5 discusses the mechanical properties of connections and Section 5.6 the forces that can appear in the connections. Then a structural check of the connection should prove, as with all other structural checks, that the design strength is greater than or equal to the forces in the connections caused by the design loads.

5.1 MECHANICAL FASTENERS

Fasteners for sections, sheets, and sandwich panels are discussed separately. While some fasteners are specific for certain cases, others can be used in all applications.

1 Mechanical Fasteners for Sections

Table 5.1 gives a general overview of the application of the different mechanical fasteners in cold-formed sections. The *European Recommendations for Steel Construction* provides more information (ECCS, 1983b).

Bolts with Nuts. Bolts with nuts are threaded fasteners which are assembled in preformed holes through the material elements to be joined. Thin members will necessitate the use of bolts threaded close to the head. Head shapes may be hexagonal, cup, countersunk, or hexagon flanged (Fig. 5.1). The nuts should normally be hexagonal. For thin-walled sections bolt diameters are usually M5 to M16. The preferred property classes are 8.8 or 10.9 according to ISO 898/I (1987).

Screws. The two main types of screws are self-tapping and self-drilling. Most screws will be combined with washers to improve the load-bearing capacity of the fastening or to make the fastening self-sealing. Some types have plastic heads or

Table 5.1 General overview of application field for mechanical fasteners

Thin to thick	Steel to wood	Thin to thin	Fastener	Remarks
×		×	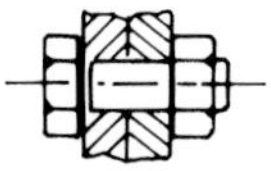	Bolts M5-M16
×				Selftapping screw ø 6.3 with washer ≥ ø 16 mm, 1 mm thick with elastomer
	×	×		Hexagon head screw ø 6.3 or 6.5 with washer ≥ ø 16 mm, 1 mm thick with elastomer
×		×	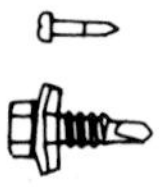	Self-drilling screw with diameters: ø 4.22 or 4.8 mm, ø 5.5 mm, ø 6.3 mm
×				Thread cutting screw ø 8 mm with washer ≥ ø 16 mm, 1 mm thick with or without elastomer
	×		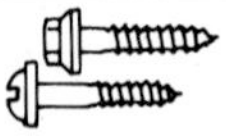	Wood-screws ø 6 mm with washer ≥ ø 16 mm, 1 mm thick
		×		Blind rivets with diameters: ø 4.0 mm, ø 4.8 mm, ø 6.4 mm
×			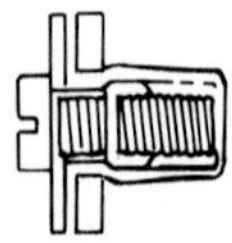	Nuts

plastic caps, which are available for additional corrosion resistance and color matching.

Self-Tapping Screws. Self-tapping screws tap their counterthread in a prepared hole. They can be classified as thread-forming and thread-cutting.

Figure 5.2 shows the thread types for thread-forming screws. Type A is used for fastening thin sheets to thin sheets, Type B for fixing to steel bases of thicknesses greater than 2 mm (0.08 in.). Type C is generally used for fixing to thin steel bases up to 4 mm (0.16 in.) thick.

Thread-forming screws normally are fabricated from carbon steel (plated with zinc for corrosion protection and lubrication) or stainless steel (plated with zinc for lubrication).

Figure 5.3 shows some examples of threads and points of thread-cutting screws. Thread-cutting screws have threads of machine screw diameter-pitch combinations with blunt points and tapered entering threads having one or more cutting edges and chip cavities. They are used for fastening to thicker metal

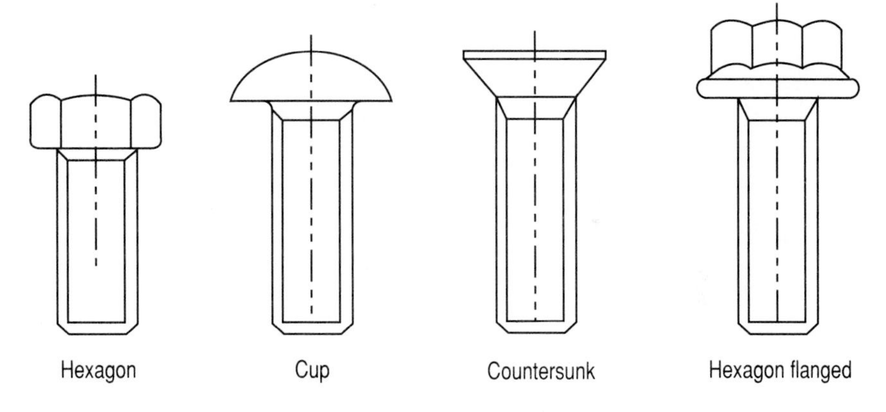

Fig. 5.1 Bolt head shapes.

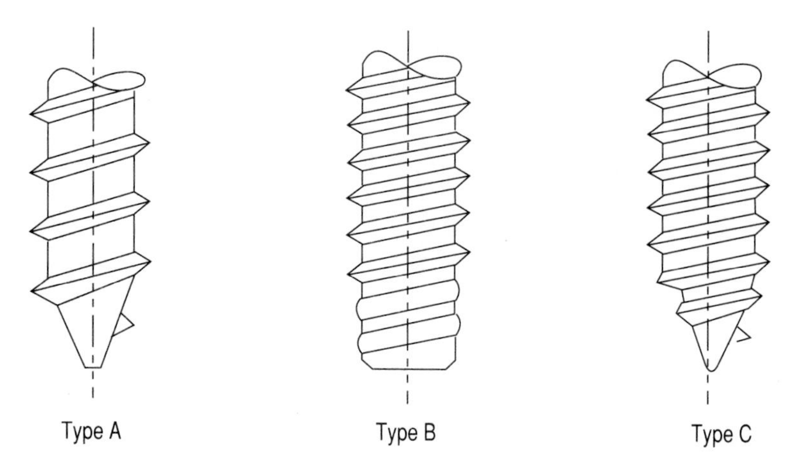

Fig. 5.2 Thread types for thread-forming screws.

bases. Resistance to loosening is normally not as high for thread-cutting screws as for thread-forming screws.

Thread-cutting screws are fabricated from carbon steel case-hardened and normally plated with zinc for corrosion protection and lubrication.

Self-Drilling Screws. Self-drilling screws drill their own holes and form their mating threads in one operation. Figure 5.4 shows two examples of self-drilling screws. The screw in Fig. 5.4*b* serves to fasten thin sheets to thin sheets.

Self-drilling screws are normally fabricated from heat-treated carbon steel (plated with zinc for corrosion protection and lubrication) or from stainless steel (with carbon-steel drill point and plated with zinc for lubrication).

Blind Rivets. A blind rivet is a mechanical fastener capable of joining workpieces together where access to the assembly is limited to one side only. Typically blind rivets are set by actuation of a self-contained mechanical feature

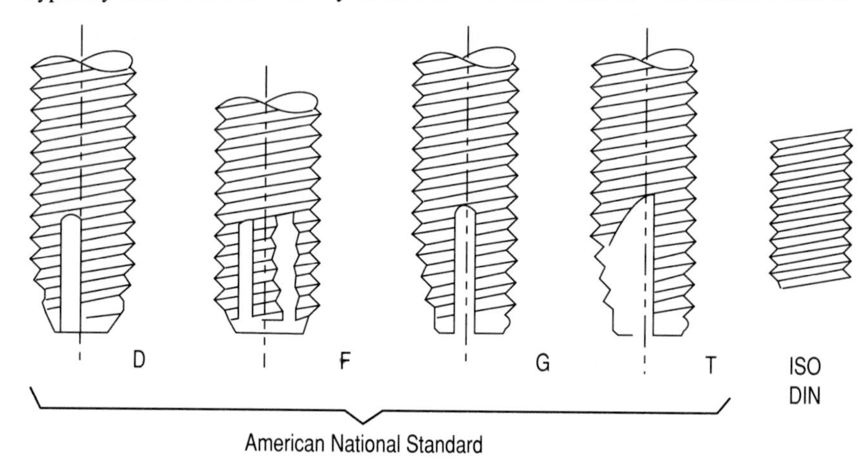

Fig. 5.3 Threads and points of thread-cutting screws.

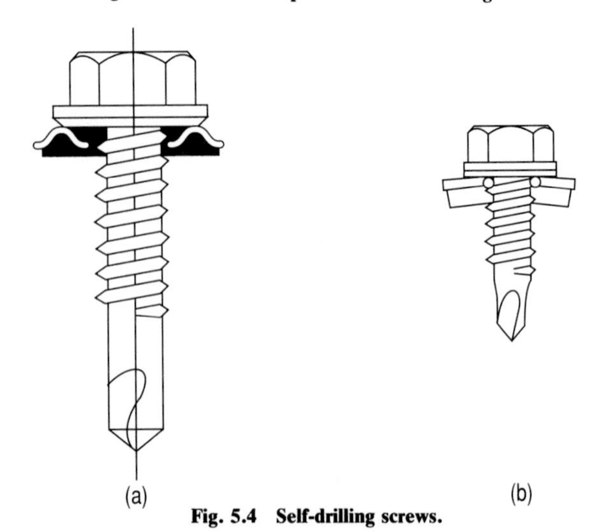

Fig. 5.4 Self-drilling screws.

which forms an upset on the blind end of the rivet and expands the rivet shank. The rivets are installed in predrilled holes. They are used for thin to thin fastening.

Blind rivets are available in aluminum alloy, Monel (nickel-copper alloy), carbon steel, stainless steel, and copper alloy. Figure 5.5 shows different types of blind rivets.

Nuts. A number of systems are available where a nut will be fastened to one of the parts to be connected. They can be used when loose nuts cannot be applied or when it is not possible to make a sufficiently strong fastening with screws. The thickness of the parts to be fastened will not limit the application. Figure 5.6 shows a number of examples.

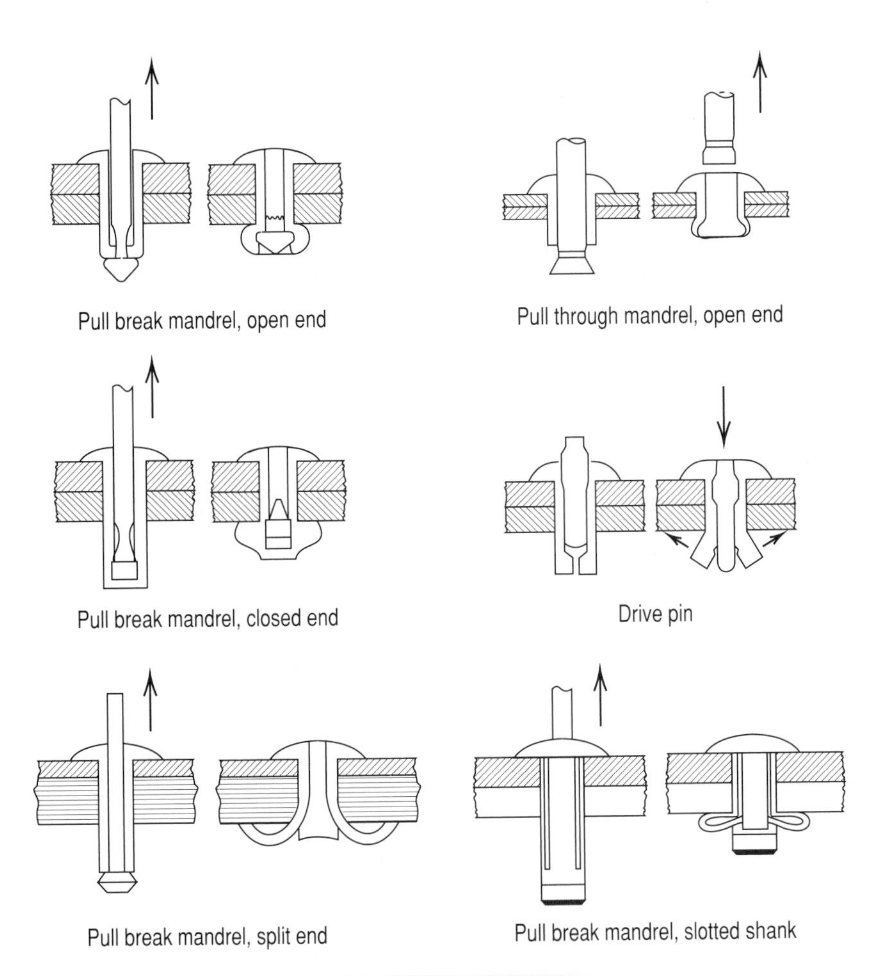

Fig. 5.5 Blind rivets.

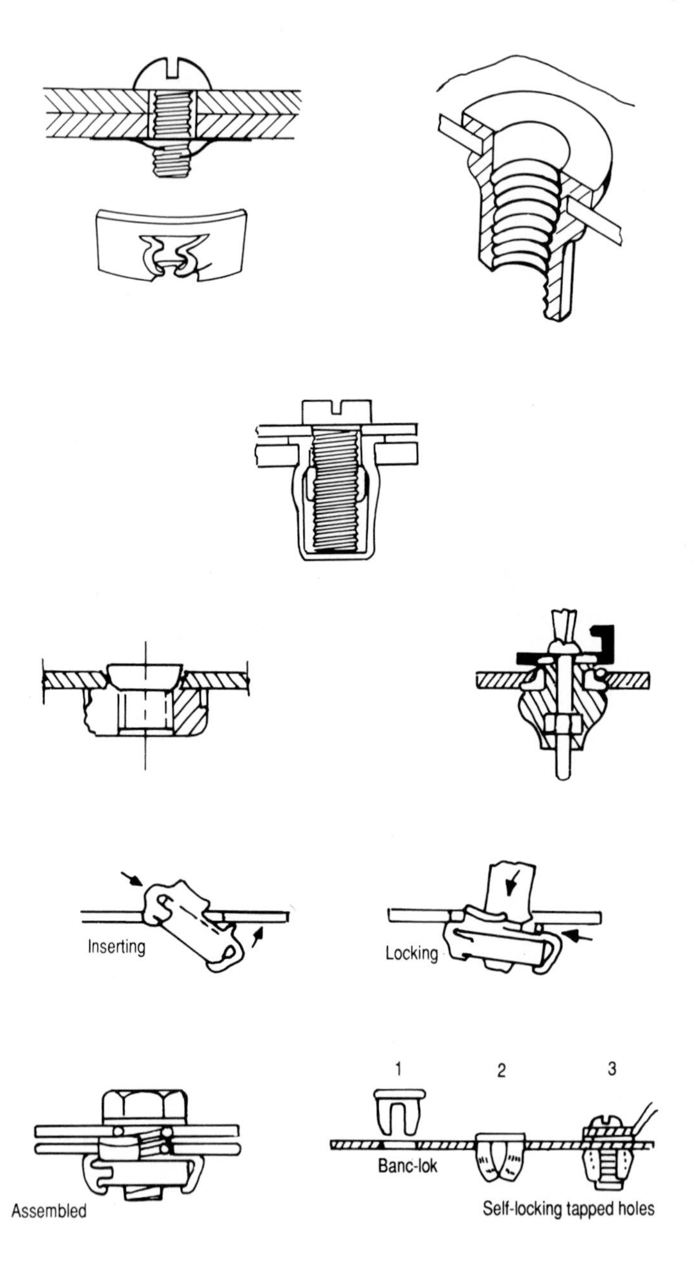

Fig. 5.6 Nut systems.

2 Mechanical Fasteners for Sheeting

In sheeting applications screws are used for fastening to the substructure, and screws and blind rivets are used for fastening at the seams. Furthermore shot pins and seam locking are also applied.

Shot Pins. Shot pins are fasteners driven through the material to be fastened into the base metal structure. Depending on the type of driving energy, there are:

Powder-actuated fasteners, which are placed with tools containing cartridges filled with propellant that will be ignited [minimum thickness of substructure 4 mm (0.16 in.)]

Air-driven fasteners, which are placed with tools that act on compressed air [minimum thickness of substructure 3 mm (0.12 in.)]

Figure 5.7 shows examples of shot pins.

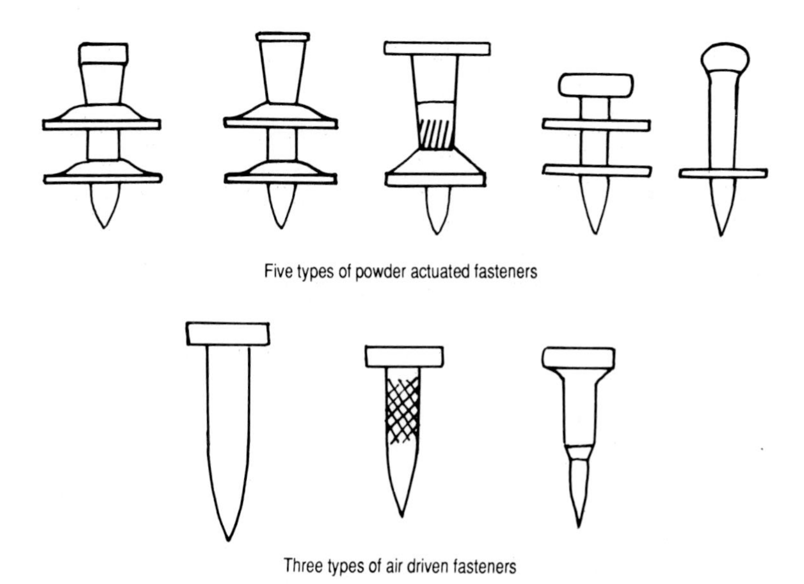

Fig. 5.7 Shot pins.

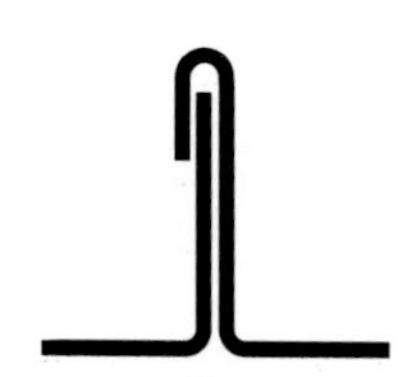

Fig. 5.8 Seam locking in sheeting.

Seam Locking. Seam locking is used in structural applications as longitudinal connection between adjacent sheets. Figure 5.8 illustrates seam locking in sheeting.

3 Mechanical Fasteners for Sandwich Panels

For sandwich panels the following fastening principles are available:

1. Fastening over the width of the panel. Fastening through the entire panel is most frequently made with screws or bolts. For fastening to the skin which is in contact with the support member, it may be necessary to strengthen the panel. The design must ensure that delamination will not occur.
2. For fastening the panel at the spot of the seam, special systems have to be designed, such as fastening through the seam and clamping over the outer skin, or fastening the inner skin by clamping it. The design of the fastening at the inner skin in the seam must ensure that delamination will not occur. One also has to be aware of a special type of delamination—peeling.
3. Special designs are possible, which may not be covered by principle 1 or 2. An example of a special design is fastening with an oversized hole to allow extension of the panel caused by temperature changes.

For flashings and covers of seams, blind rivets or screws are often being used.

5.2 WELDING

In general, cold-formed sections are suitable for welding, and joints can be made by the open-arc process as well as by resistance welding.

1 Open-Arc Welding

The principle of open-arc welding is a heat input by an electric open arc which is being struck between an electrode and a workpiece. Either covered electrodes or welded wires are used as welding consumables. During the welding process the molten pool is protected by slag or gas against harmful environmental influences.

The following welding procedures are normally used for thin-walled sections:

1. *GMA (gas-metal arc) welding.* An open arc burns between the wire and the base material. For protection, shielding gases of CO_2 as well as mixed gases are used.
2. *Manual arc welding (welding with covered electrodes).* The open arc burns between the electrode and the base material. The welding characteristics and the quality of the weld metal are strongly influenced by the type of covering.
3. *TIG (tungsten–inert gas) welding.* The open arc is struck between a tungsten electrode and the workpiece. Argon is the shielding gas which protects the welding process. The welding consumable is added separately as either wire or rod.
4. *Plasma welding.* With this procedure a plasma is produced between a tungsten electrode and the base material. Compared with the TIG process, the energy input during the welding procedure is more concentrated. Figure 5.9 shows a diagram of the plasma process.

2 *Resistance Welding*

Resistance welding is done without an open arc. There is no need for protection of the molten metal by shielding gas or slag. During this procedure special electrodes lead the welding current with high density to the base material. The workpiece is being so strongly heated locally that it turns into a plastic condition and then starts to melt. In this condition, pressure transferred by the electrodes to the workpiece leads to a local connection of the construction pieces. The processes used include spot welding, roller seam welding, and projection welding. Figure 5.10 illustrates the resistance-welding procedures.

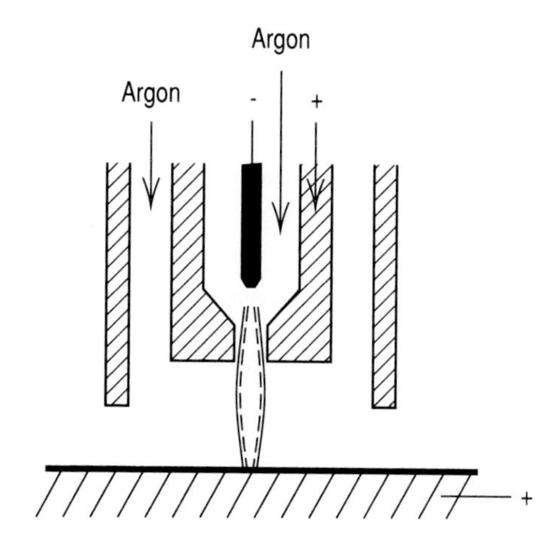

Fig. 5.9 Plasma welding process.

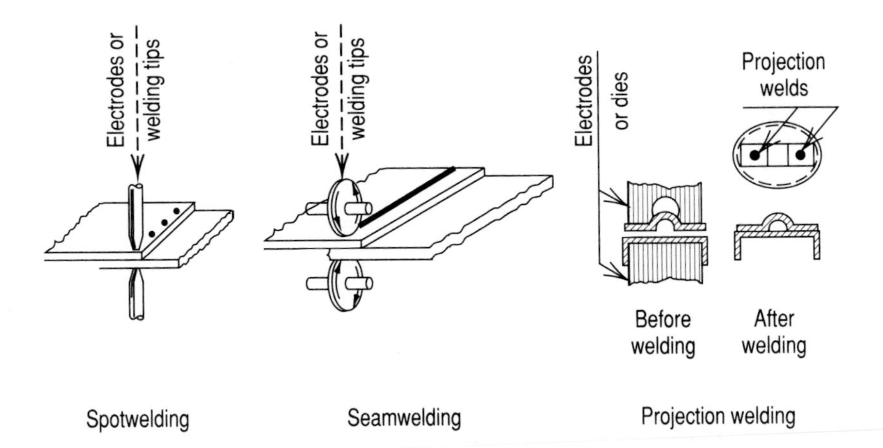

Fig. 5.10 Spot, seam, and projection welding processes.

Table 5.2 Requirements for connections in thin-walled structures

Structural requirements
1. Strength 2. Stiffness 3. Deformation capacity
Nonstructural requirements
1. Economic aspects a. Total number of fastenings to be made b. Skill required c. Ability to be dismantled d. Design life e. Installation costs Piece cost of fastener Direct labor Indirect labor Application tools Maintenance Inventory 2. Durability a. Chemical aggressiveness of environment b. Possible galvanic corrosion 3. Watertightness 4. Aesthetics

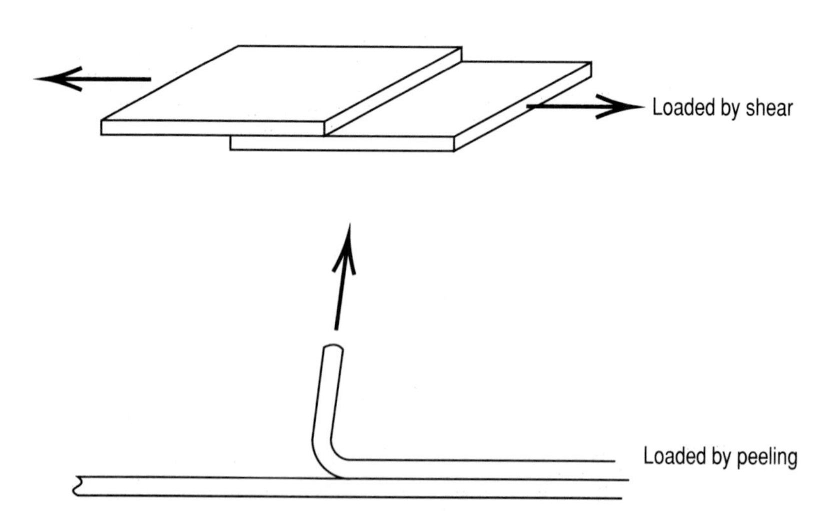

Fig. 5.11 Shear and peeling of connections made by adhesive bonding.

5.3 ADHESIVE BONDING

For fastening by means of bonding, it is important to realize that a bonded connection possesses a good shear resistance but mostly a bad peeling resistance (Fig. 5.11). For that reason a combination of bonding and mechanical fastening is sometimes chosen.

Adhesives used for thin-walled steels are epoxy types, in which the best hardening will appear under elevated temperature [on the order of 80 to 120°C (176 to 248°F)], and acrylic types, which are more flexible than the epoxy types.

Two advantages of bonded connections are a uniform distribution of forces over the connection and good repeated load behavior. Some disadvantages are that the surfaces should be flat and clear and there is a hardening time.

5.4 REQUIREMENTS AND SELECTION PROCEDURE

Connections are an important part of every structure, not only from the point of view of structural behavior, but also in relation to the cost of production. For economic reasons, the influence of the joining process on cost will tend to increase.

A variety of joining methods are available for lightweight structures. Correct selection is governed by a large number of factors (see Table 5.2).

The choice of the type of fastening will primarily be governed by the nonstructural requirements. Then the number of fasteners or area of fastening will be determined by the structural requirements, which include a check of the mechanical properties of the connections compared with the possible forces in the connections.

5.5 MECHANICAL PROPERTIES OF CONNECTIONS

1 General

The important mechanical properties of connections are strength (capacity), stiffness, and deformation capacity.

Strength. The strength of connections is dependent on the type of fastener and the properties of the jointed elements (thickness, yield stress). The most reliable strength values are determined by testing in accordance with the *European Recommendations for Steel Construction* (ECCS, 1983a). Later in this section, formulas are provided (in accordance with Eurocode 3 Annex A, 1991) for the characteristic shear or tension strengths per fastener type. These formulas lead to values which are on the conservative side. The application limits for those formulas are given in accordance with Eurocode 3 Annex A (1991). The material factor for these connections should be taken as $\gamma_m = 1.25$. (See Bryan et al., 1990, which provides background information.)

For rather small thicknesses of steel (such as trapezoidal sheeting) the tension-loaded connections are sensitive to repeated loads. In Eurocode 3 Annex

A (1991) this has been taken into account by an additional factor of 2 at the static strength when the connection is loaded by a spectrum comparable to wind load. In ECCS (1991) testing methods and design formulas are given for connections in sandwich panels.

Stiffness. The stiffness of connections is important because it determines the stiffness of the entire structure or its components. Moreover, the stiffness of the connections will influence the force distribution within the structure. Especially when the connection is a part of a bracing structure, the stiffer the connection, the lower the bracing force will be. With cold-formed sections, special systems are available where the sections interlock to form connections with good stiffness.

Deformation Capacity. The deformation capacity of a fastening is important with regard to good force distribution in the connection and the structure. A connection with no deformation capacity can cause the brittle fracture of a structure or element.

2 Strength Formulas and Application Limits for Connections in Accordance with Eurocode 3 Annex A (1991)

Strength Formulas for Connections with Mechanical Fasteners. In addition to determining strength characteristics by testing, appropriate formulas can also be used. For fastenings under a single static load, the characteristic shear strength can be determined according to Table 5.3, where the values given are on the safe side. Table 5.4 lists the values for fastenings loaded in tension.

Symbols Used in Tables 5.3 and 5.4

A_{et} = tensile stress area of bolt
A_n = net cross-sectional area of plate material
d = nominal diameter of hole
d_n = nominal diameter of fastener
d_w = diameter of washer or head of fastener
e_1 = edge distance in load direction
e_2 = center-to-center distance in load direction
F_b^* = characteristic shear strength of a connection per fastener, failure-mode hole bearing (including tilting and shear of section)
F_n^* = characteristic strength of a connection, failure-mode yield of net section
F_o^* = characteristic pullout strength of a connection per fastener
F_p^* = characteristic pull-through–pull-over strength of a connection per fastener
F_v^* = characteristic capacity per bolt per shear plane (for thin-walled steel, thread is always up to head)
f_n = design value for net section stress
f_u = specified ultimate tensile strength of steel sheet with thickness t
f_{ub} = specified ultimate tensile strength of bolts
r = force transmitted by bolt or bolts at section considered, divided by tension force in member at that section
t = thickness of thinnest sheet
t_1 = thickness of thickest sheet (base material)
u_1 = distance between edge and center of fastener perpendicular to load direction
u_2 = center-to-center spacing of fasteners perpendicular to load direction
α = factor defined at relevant place

Table 5.3 Characteristic shear strengths for fastenings

Fastener type	Failure mode: Hole bearing (including tilting) 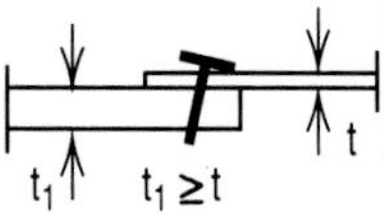	Failure mode: Failure of net section 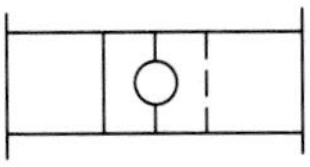	Failure mode: Shear of fastener 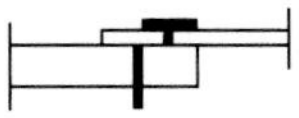	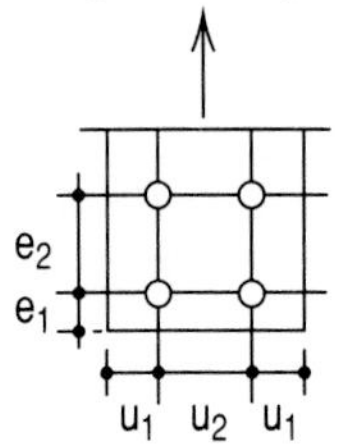 Range of validity†
	t_1, $t_1 \geq t$, t			e_2, e_1, u_1, u_2, u_1
Blind rivets	$F_b^* = \alpha f_u d_n t$ For $t = t_1$, $\alpha = 3.6(t/d_n)^{1/2} \leq 2.1$ For $t_1 \geq 2.5t$, $\alpha = 2.1$ For $1 < t_1/t < 2.5$, linear interpolation	$F_n^* = A_n f_u$	Characteristic shear strength of fastener should be 1.2 times larger than for other failure modes	$e_1 \geq 3d_n$, $e_2 \geq 3d_n$ $u_2 \geq 3d_n$, $u_1 \geq 1.5d_n$ $2.6\ \text{mm} \leq d_n \leq 6.4\ \text{mm}$
Bolts with nuts	$F_b^* = 2.5\alpha t d_n f_u$ where α is the lesser of $e_1/3d_n$ or 1	$F_n^* = A_n f_n$ where $f_n = (1 - 0.9r + 3rd/u)f_u$ with a maximum of $f_n = f_u$ u is the lesser of $2u_1$ or u_2	$F_v^* = 0.6 f_{ub} A_{et}$ for $f_{ub} \leq 800\ \text{N/mm}^2$ $F_v^* = 0.5 f_{ub} A_{et}$ for $f_{ub} > 800\ \text{N/mm}^2$ F_v^* should be 1.2 times larger than for other failure modes	$e_1 \geq 1.5d_n$, $e_2 \geq 3d_n$ $u_2 \geq 3d_n$, $u_1 \geq 1.5d_n$ Minimum bolt size M6 Bolt classes 4.6 to 10.9 $t \geq 1.25\ \text{mm}$
Cartridge fired pins	$F_b^* = 3.2 f_u d_n t$	$F_n^* = A_n f_u$	Characteristic pullout load due to shear shall exceed F_b^* by 50%	$e_1 \geq 4.5d_n$, $e_2 \geq 4.5d_n$ $u_2 \geq 4.5d_n$, $u_1 \geq 4.5d_n$ $3.7\ \text{mm} \leq d_n \leq 6.0\ \text{mm}$ $d_n = 3.7\ \text{mm} \rightarrow t_1 \geq 4\ \text{mm}$ $d_n = 4.5\ \text{mm} \rightarrow t_1 \geq 6\ \text{mm}$ $d_n = 5.2\ \text{mm} \rightarrow t_1 \geq 8\ \text{mm}$
Screws	$F_b^* = \alpha f_u d_n t$ For $t = t_1$, $\alpha = 3.2(t/d_n)^{1/2} \leq 2.1$ For $t_1 \geq 2.5t$, $\alpha = 2.1$ For $1 < t_1/t < 2.5$, linear interpolation	$F_n^* = A_n f_u$	Characteristic shear strength of fastener should be 1.2 times larger than for other failure modes	$e_1 \geq 3d_n$, $e_2 \geq 3d_n$ $u_2 \geq 3d_n$, $u_1 \geq 1.5d_n$ $3.0\ \text{mm} \leq d_n \leq 8.0\ \text{mm}$

† Fasteners may be used beyond these limits, but the characteristic strength should then be based only on experimental evidence.

Combination of Shear and Tension Load. When a fastening is subjected simultaneously to shear and tension, it should be verified that

$$\frac{F_t}{F_{dt}} + \frac{F_s}{F_{ds}} \leq 1 \tag{5.1}$$

where F = force in a fastening caused by design load
F_d = design strength of a fastening

and the subscripts t and s indicate tension and shear, respectively.

Strength Formulas for Welded Connections. The provisions in this section apply to (1) spot-welded sheet steel, resistance- and fusion-welded; and (2) lap fillet

Table 5.4 Characteristic tension strengths for fastenings

Fastener type	Failure mode: Pull-through, pull-over†	Failure mode: Pullout	Failure mode: Tensile failure	Application limit for formulas
Screws	$F_p^* = d_w t f_u$ or $F_p^* = 0.5 d_w t f_u$	$F_o^* = 0.65 t_1 d_n f_{uc}$‡	Characteristic tensile strength of fastener $> F_p$ or F_o¶	$0.5\ \text{mm} < t < 1.5\ \text{mm}$; $t_1 > 0.9\ \text{mm}$
Cartridge-fired pins	$F_p^* = d_w t f_u$ or $F_p^* = 0.5 d_w t f_u$	Characteristic pullout strength $> F_p$‖	Not relevant in sheeting	$0.5\ \text{mm} < t < 1.5\ \text{mm}$; $t_1 > 6\ \text{mm}$

† (1) $F_p^* = d_w t f_u$ applies for static loads; $F_p^* = 0.5 d_w t f_u$ applies for repeated loads with a spectrum similar to wind. (2) The failure mode "gross distortion" is not covered by the given formulas. For fastenings through flanges having a width smaller than 150 mm (6 in.) in the local deformation of the flange under working load is in most cases in the elastic range. (3) It is assumed that load is applied centrally and the washer has sufficient rigidity to prevent it from being deformed appreciably or pulled over the head of the fastener. When attachment is at a quarter point, the design value is $0.9F_p^*$; and when it is at both quarter points, the design value is $0.7F_p^*$.

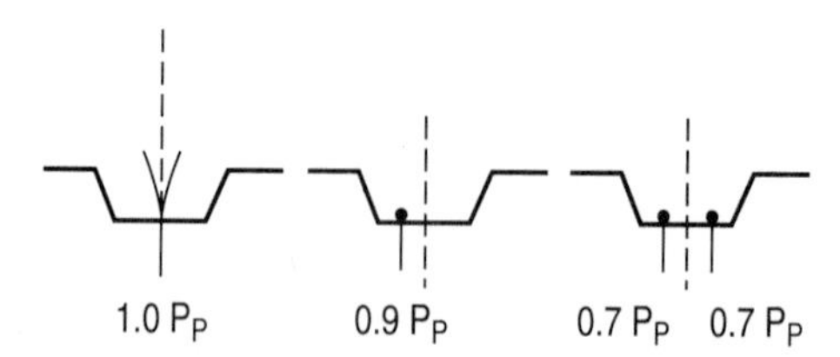

‡ When $F_o^* \ll F_p^*$, it has to be proven that the deformation capacity is sufficient.

¶ For more than one sheet layer the tensile strength of the fastener should be a multiple of F_p^* or F_o^*. The factor depends on the number of sheet layers.

Remark. For the diameter of predrilled holes for screws, the manufacturer's guidelines shall be observed. These guidelines should be based on the following criteria:

1. The applied torque should be a little higher than the threading torque.
2. The applied torque should be lower than the thread-stripping torque or head-shearing torque.
3. The threading torque should be smaller than $^2/_3$ of the head-shearing torque.

welds, fusion-welded. The clause applies for welds in parent materials with thicknesses equal to or smaller than 4 mm (0.16 in.).

Spot-Welded Sheet Steel Connections. The characteristic shear capacities are given in Table 5.5.

Symbols Used in Table 5.5

A_n = net cross-sectional area of plate material, mm^2
d_s = diameter of spot weld, mm (for fusion welding, $d_s = 5 + 0.5t$; for resistance welding, $d_s = 5t^{1/2}$)
e = end distance of spot weld, mm
F^* = characteristic shear capacity of single spot weld, N (subscript defines failure mode)
f_u = ultimate tensile strength of material, N/mm^2
t = thickness of thinnest connected part, mm
t_1 = thickness of thickest connected part, mm

Depending on the weld procedure, the value of d_s is controlled by shear stress tests of single-lap connections constructed according to Fig. 5.12. The characteristic shear capacity R is determined by statistical evaluation of the test results. The requirement is that

$$R > F_s^* \tag{5.2}$$

Constructional Details of Spot-Welded Sheet Steel Connections

1. *Spacing of spot welds.* The end distance from a spot weld to the edge of the connected part in the load direction should be not less than $2d_s$ and not more than $6d_s$. The edge distance at right angles to the load direction should not exceed $4d_s$. The center-to-center spacing of spot welds should be not less than $3d_s$ and not more than $8d_s$ in the load direction; it should not be more than $6d_s$ at right angles to the load direction.
2. *Maximum thickness of sheet.* The thickness of the thinnest sheet to be spot-welded should not be greater than 3.0 mm (0.12 in.).

Lap Fillet Welds. The characteristic shear capacity is given by the following

Table 5.5 Characteristic shear capacities per spot weld

Failure mode				
Shear of spot weld*	Tearing and bearing at contour weld†	Edge failure (shear section)	Net section	Range of validity
$F_s^* = (\pi/4)d_s^2 f_u$	$F_t^* = 2.7t^{1/2} d_s f_u$	$F_e^* = 1.4tef_u$	$F_n^* = A_n f_u$	Constructional details discussed later in this section

* Necessary condition to be fulfilled: $F_s^* \leq 1.25\ F_t^*$, F_e^*, or F_n^*.

† The value of F_t^* is valid for as-rolled and hot-dipped galvanized material and for $t \leq t_1 \leq 2.5t$. If $t_1 > 2.5t$, then F_t^* should be taken as the value found in the table provided that $F_t^* \leq 0.691 d_s^2 f_u$ and $F_t^* \leq 3.08 t d_s f_u$.

formulas. For end fillet welds,

$$F^* = tl_w f_u\left(1 - 0.3\frac{l_w}{b}\right) \quad \text{per weld} \tag{5.3}$$

and for side fillet welds,

$$F^* = 2tl_w f_u\left(0.9 - 0.45\frac{l_w}{b}\right) \quad \text{per two welds,} \tag{5.4}$$

provided that

$$l_w < b \tag{5.5}$$

where b = width of specimen
l_w = effective length
Other symbols were defined earlier.

The effective length of a fillet weld shall be the overall length, including end returns. No reduction in effective length shall be made for either the start or the termination of the weld. Weld lengths shorter than eight times the thickness of the thinnest connected sheet should be ignored for transmission of forces.

When, in a joint, a combination of end fillet welds and side fillet welds is used, the total capacity shall be the sum of the capacities of the individual welds.

Constructional Details of Lap Fillet Welds. The welding parameters should be chosen such that the capacity is governed by the thickness of the sheet. This will be achieved when the smallest cross section of the weld is at least equal to the cross section of the connected sheet.

Circular Plug Welds. Circular plug welds should not be designed to transmit any forces other than shear. They should not be used where the thinnest connected part is over 4 mm (0.16 in.) thick or through connected material having a total thickness of more than 4 mm (0.16 in.). Weld washers should be used when the thickness of the connected material is less than 0.7 mm (0.03 in.). Circular plug welds should have a minimum effective diameter of 10 mm (0.4 in.).

The characteristic shear capacity F_s^*, in newtons, for circular plug welds may

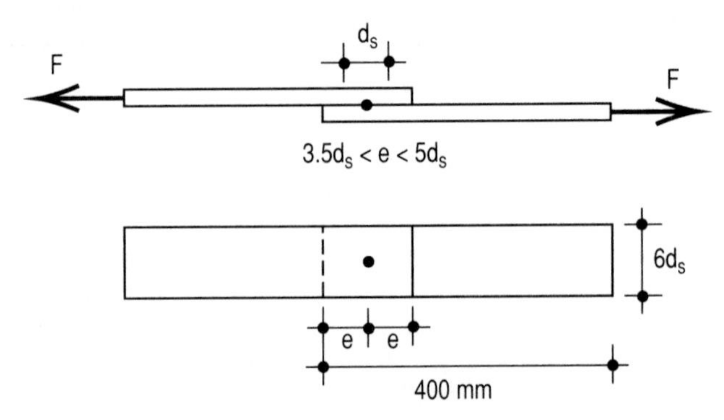

Fig. 5.12 Shear failure test specimen.

be taken as the smaller of case 1 or 2:

1. For $\frac{d_{\text{eff}}}{t} \leq 22\left(\frac{280}{f_y}\right)^{1/2}$,

$$F_s^* = 1.33td_{\text{eff}}f_u \tag{5.6}$$

For $22\left(\frac{280}{f_y}\right)^{1/2} < \frac{d_{\text{eff}}}{t} < 38\left(\frac{280}{f_y}\right)^{1/2}$,

$$F_s^* = 0.17\left[1 + \frac{150t}{d_{\text{eff}}}\left(\frac{280}{f_y}\right)^{1/2}\right]td_{\text{eff}}f_u \tag{5.7}$$

For $\frac{d_{\text{eff}}}{t} \geq 38\left(\frac{280}{f_y}\right)^{1/2}$,

$$F_s^* = 0.84td_{\text{eff}}f_u \tag{5.8}$$

2. For all values of d_{eff}/t,

$$F_s^* = 0.38(0.7d_w - 1.5t)^2 f_w \tag{5.9}$$

where d_{eff} = effective diameter of weld, mm
d_w = visible diameter of weld, mm
f_u = design strength of steel, N/mm^2
f_w = design strength of weld material, N/mm^2
f_y = yield strength of steel, N/mm^2
t = thickness of plate material, mm

Elongated Plug Welds. The general recommendations for elongated plug welds are the same as those for circular plug welds.

The characteristic shear capacity F_s^*, in newtons, for elongated plug welds may be taken as the smaller of the following equations:

$$F_s^* = (0.4L_w + 1.33d_{\text{eff}})tf_u \tag{5.10}$$

$$F_s^* = [(0.43d_w - 0.92t)^2 + L_w(0.35d_w - 0.75t)]f_u \tag{5.11}$$

where L_w is the length of the weld, mm, and d_{eff}, t, and f_u are as defined earlier.

5.6 FORCES IN CONNECTIONS

The forces in connections are caused by loads that have been defined in loading specifications and in Council on Tall Buildings, Group CL (1980). The loads cause shear or tension, or a combination of these, in the connections. The values of the forces per fastener are dependent on (1) the loads on the jointed elements, (2) the stiffness of the jointed elements, and (3) the stiffness and deformation capacity of the fastenings. Within the possible forces, the following can be distinguished:

1. *Primary forces.* Forces directly caused by the load
2. *Secondary forces.* Forces indirectly caused by the load and which may be neglected in the presence of sufficient deformation capacity in the fastening

As examples of the forces in thin-walled sections, three types of structures are presented.

1. Composition of a bending member from single sections, with connections loaded in shear (Fig. 5.13).
2. Composition of a bending member from single sections, with connections loaded in tension. Figure 5.14 shows the cross section of two C sections connected to each other and gives the method of calculating the forces in the fastenings.
3. Secondary forces in connections. Care should be taken through suitable detailing that second-order effects caused by the deformation of thin-walled sections will not generate impermissible additional forces in the fastenings. This is illustrated in Fig. 5.15.

The forces in profiled sheeting can be shear forces and tension forces. Shear forces include the following:

1. The dead weight of steel sheets, such as the weight of wall or facade elements.
2. The diaphragm action when the diaphragm is used deliberately, as in the absence of wind bracing or stability support for beams or columns.

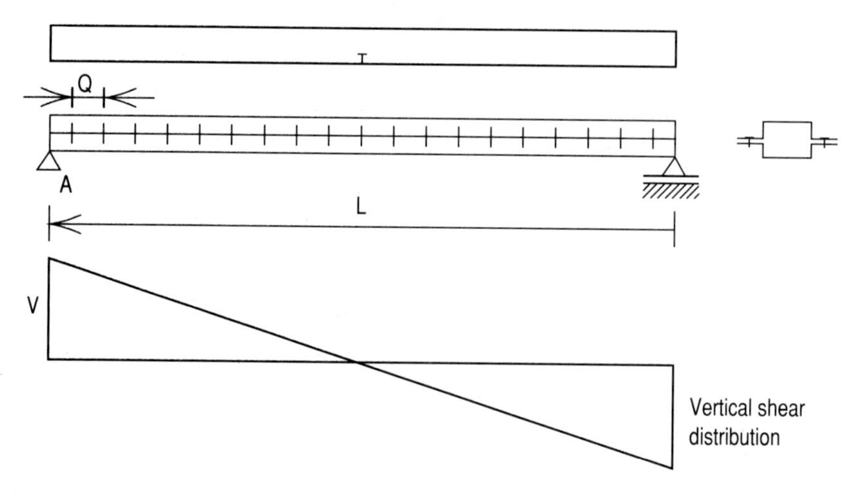

Fig. 5.13 Shear forces in composite beam.

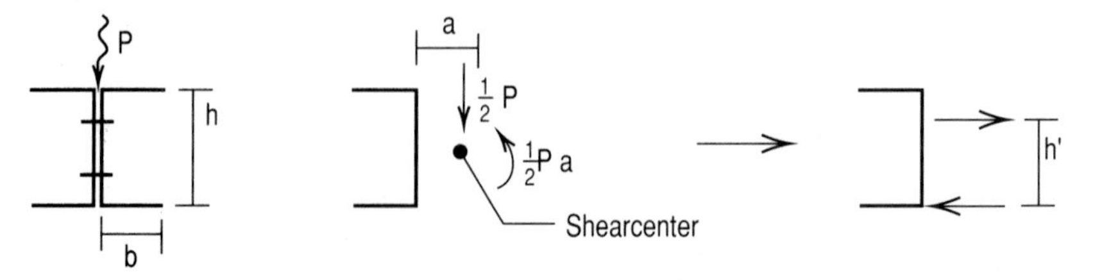

Fig. 5.14 Cross section of two C sections connected to each other. Tension force in fastener = $\frac{1}{2}Pa/h$ for $a = 3b^2/(6b + h)$.

3. Variation of the temperature of the steel sheets. With sufficient deformation capacity, the shear forces will be small and may be neglected.
4. Rotation of eccentrically fastened sheet ends and membrane action of the sheet (Fig. 5.16). If sufficient deformation capacity is provided, the fastening will not fail.
5. Diaphragm action that is not used structurally. This is the case when a sheeting or cladding is only used as an outer skin. It is then necessary for the skin to follow the deformations of the substructure. This is possible when the diaphragm (especially the fastenings) possesses sufficient deformation capacity.

Tension forces are caused mainly by loads perpendicular to the plane of the steel sheets. When determining required strength and stiffness of the sheets, a simply supported static system is assumed (Fig. 5.17). In reality the sheets are

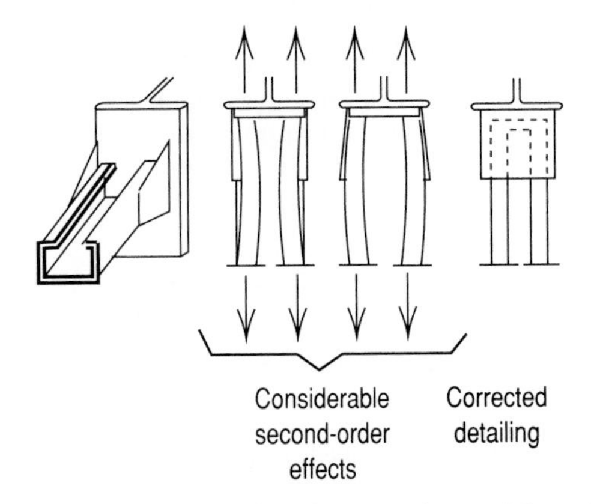

Fig. 5.15 Influence of detailing of a connection on deformations.

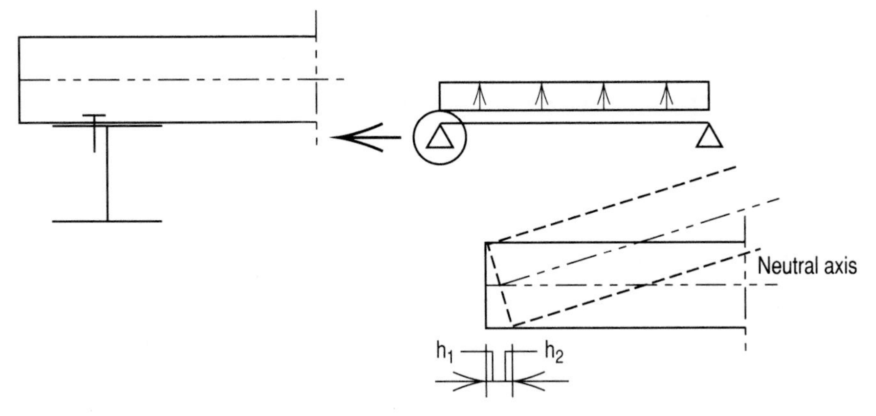

Fig. 5.16 Details of support of sheet. h_1—deflection caused by difference in length of straight and curved neutral axes; h_2—deflection caused by distance between neutral axis and flange through which it is fastened.

restrained to some extent at the supports, but for the design of the sheets it is safe to neglect the restraining effect. For the design of the fastenings, simplification to a simply supported static system results in forces that are too small. In the absence of sufficient deformation capacity, the fastening can fail at a premature stage.

The action of the internal forces in a fastening is illustrated in Fig. 5.18. Due to bending of the steel sheet, a compression force will act on the supports at point *A* or *B*. This causes an accidental fixing moment for the steel sheets, which generates an extra tension force in the fastener, the prying force. The value of the prying force depends on:

1. Stiffness of the sheets in relation to the span
2. Flexibility of the sheets near the fastener
3. Diameter of the head of the fastener or diameter and stiffness of the washer
4. Distance between fastener and contact points *A* or *B*
5. Torsional rigidity of the support

When sufficient deformation capacity is available, the required rotation can take place and the first-order reaction can proceed.

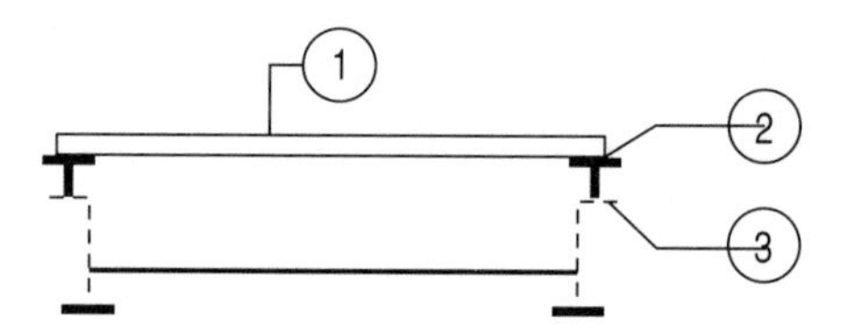

Fig. 5.17 Static system to determine required strength and stiffness of a steel sheet. ① steel sheet; ② fastener; ③ support.

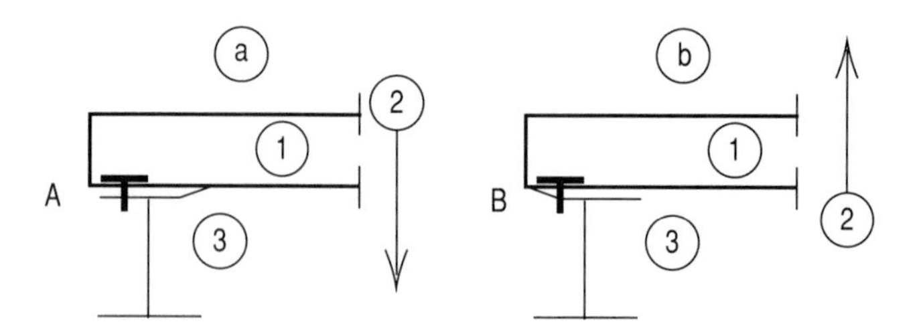

Fig. 5.18 Detail of support of a steel sheet. (*a*) downward load; (*b*) upward load. ① steel sheet; ② load; ③ support. See text for keys *A* and *B*.

5.7 CONDENSED REFERENCES/BIBLIOGRAPHY

Bryan 1990, *Evaluation of Test Results on Connections in Thin Walled Sheeting and Members in Order to Obtain Strength Functions and Suitable Model Factors*

Council on Tall Buildings, Group CL 1980, *Tall Building Criteria and Loading*

Council on Tall Buildings, Group SB 1979, *Structural Design of Tall Steel Buildings*

ECCS 1983a, *The Design and Testing of Connections in Steel Sheeting and Sections*

ECCS 1983b, *Mechanical Fasteners for Use in Steel Sheeting and Sections*

ECCS 1991, *European Recommendations for Sandwich Panels, Part I: Design*

Eurocode 3 Annex A 1991, *Cold Formed Steel Sheeting and Members*

Tomà 1982, *Connections in Thin-Walled Structures*

6

Functional Requirements for Tall Buildings

6.1 BASIC ELEMENTS

In traditional steel structural engineering, the basic element of the load-bearing structure is bar-shaped, thus involving skeleton construction, whereas using thin-walled cold-formed components, plane elements can be formed which have load-bearing and space-covering functions. This opens up a new market for steel construction in such cases where priority is given to the space-covering function and the demand for future changes in utilization is comparatively limited, for example, in residential buildings, schools, office buildings, or other structures with large surface systems such as walls and roofs in halls. Sandwich elements with plastic foam between metal skins are representative of this construction technique (see Section 3.7).

The basic concept of lightweight construction on the basis of thin-walled cold-formed building elements with load-bearing and space-covering functions is therefore oriented toward the end product and the respective functional requirements, which means that the components of the end product themselves must be functionally designed. For steel components this means optimization of the entire structure with regard to statics *and* combination with appropriate elements for the interior environment.

By adequate shaping of profiled sheets or by the addition of linear cold-formed sections, plane building elements may be developed that have the load-bearing capacity and stiffness necessary for space-covering elements. In many fields of application, these elements render the steel skeleton unnecessary, or contribute to the load-bearing action of the skeleton. If, however, this function is fulfilled by a plane building element, adequate materials can be used to satisfy the remaining functional requirements, such as weatherproofing, sound insulation, and fire protection—in other words, materials that do not have to meet requirements concerning load-bearing capacity. The characteristic feature of lightweight construction is, therefore, the reasonable combination of building materials—oriented toward the end product—provided that the load-bearing function is fulfilled by cold-formed steel sheeting. The functional requirements can be classified as follows:

1. Load-bearing capacity
2. Stiffness
3. Durability
4. Fire protection
5. Sound insulation
6. Climatic protection (weatherproofing)
7. Room environment
8. Servicing facilities

Requirements 1 to 3 have to cover the basic safety aspects, irrespective of type and utilization of the buildings, whereas the other requirements depend on the type of building and—beyond the minimum requirements—on quality demands. Minimum requirements are specified in national standards; quality-directed requirements, at the request of the user or the owner, can be considered as improvement measures and are subject to economy aspects.

Requirement 7 includes, beyond requirements 1 to 6, those qualities that are necessary for the physiological well-being of the user. A pleasing atmosphere has a considerable influence on whether or not a building technique is acceptable to the user.

Requirement 8 is of particular importance as installations are laid in vertical and horizontal zones of the structure, and it therefore exerts influence on the structural design of the flooring systems and the realization of requirements 1 to 7. (See also Council on Tall Buildings, Group SC, 1980.)

Technically these requirements are not independent of each other. The only purpose of the formal breakdown is to facilitate the analysis of appropriate measures to satisfy the requirements. Relevant references are Baehre (1978, 1982), ECCS (1984), Balasz (1980), König (1981), and Balasz and Thomasson (1985).

1 General Functional Requirements

The building elements to be treated are subsystems of the total technical system of a building and have load-bearing, space-enclosing, and servicing functions (Fig. 6.1). An essential part of the project is to satisfy the functional demands, including measures specified in technical building regulations as well as application-oriented requirements and cost guidelines. The combination of essential requirements involves the careful aesthetic optimization of the system, taking servicing requirements into account.

The problems of servicing are exemplified in Fig. 6.2 for a one-corridor office building and a residential building, where installations of various types have to be located within the available space of the floor construction. As can be seen, considerable space is required for servicing. This means in any case that either the overall height of the floor and supporting structure must be limited or necessary clearance must be available which, on the other hand, affects the structural design of the floor.

2 Requirements—Building Authorities

The requirements on the part of the building authorities are defined in national standards. As a rule, they are minimum requirements and cover:

1. *Load-bearing capacity.* Analysis of stability and serviceability under defined nominal loads.
2. *Stiffness.* Observance of deflection limitations.
3. *Durability.* Preservation of functional stability (requirements 1 and 2), taking into account the long-term behavior of the components and the load history.
4. *Fire protection.* Observance of the fire-resistance grade depending on the type of structure and the function of the building elements with regard to fire protection, taking into account material properties such as inflammability, combustibility, and development and toxicity of fumes.
5. *Sound insulation.* Observance of tolerance dimensions depending on the type of structure and the function of the building elements with regard to

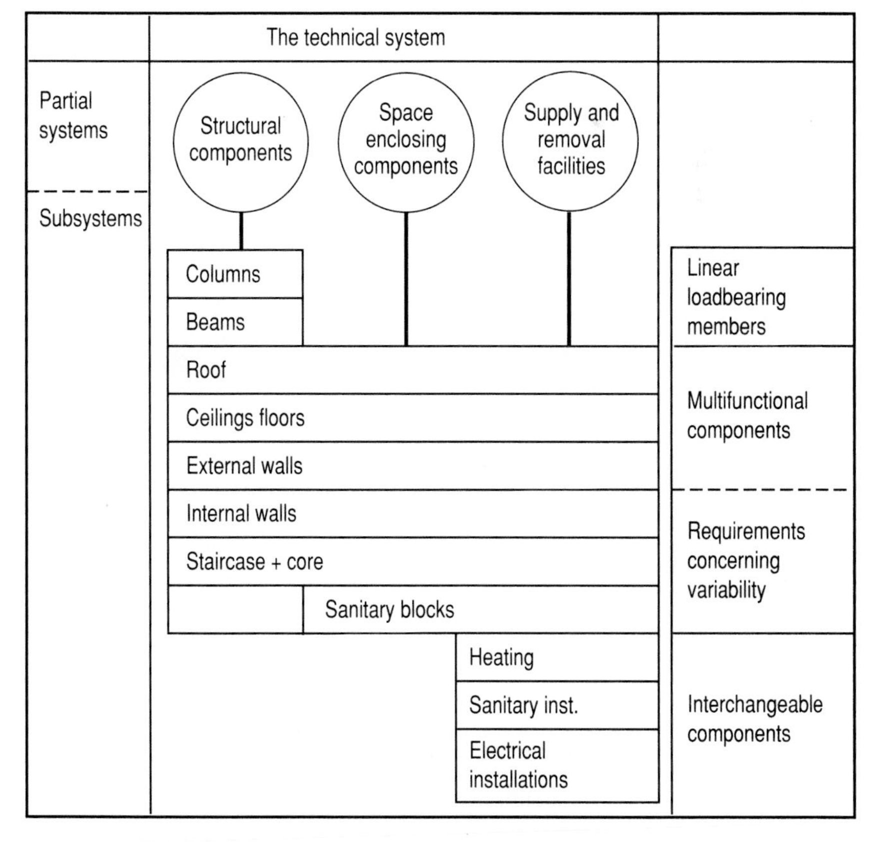

Fig. 6.1 Technical system of a building with partial systems and subsystems.

sound insulation (airborne sound, footstep sound, flanking transmission, sound absorption, and noise pollution).

6. *Climatic protection.* Adherence to thermal insulation requirements depending on climatic conditions, particularly when the building elements protect the covered space from the external climate. Problems such as vapor diffusion, air permeability, convection, and heat capacity have to be taken into consideration.

3 Requirements—Users and Owners

The minimum requirements on the part of the users are covered by the building authorities' regulations. Increased demands may be the result of special utilization requirements and are then part of a contract with the owner. They can be improvements or special solutions concerning stiffness (2), sound insulation (5), climatic protection (6), room environment (7), or servicing (8). A carefully designed building system should allow for such functional improvements or modifications.

Fig. 6.2 Illustration of servicing demands.

4 Consequences and Conclusions

The total demand on a building system is the sum of:

1. *Minimum requirements,* in part generally valid, in part building-dependent, defined in technical building regulations, *and*
2. *Additional demands,* quality- or value-improving measures at the user's or owner's request.

It is therefore essential that a basic system be available which fulfills the minimum requirements and can be modified by appropriate measures to satisfy additional quality demands.

As the type and the extent of minimum requirements also depend on the types of structure and components, it is advisable to prepare application-oriented quality profiles defining the minimum and additional requirements qualitatively and quantitatively and to use them as a basis for product development. The following section gives examples of such quality profiles.

6.2 SPECIFIC QUALITY REQUIREMENTS (QUALITY PROFILES)

The specific functional requirements of three selected building elements are exemplified in this section. As international regulations are not uniformly stringent, the requirements are presented on a six-grade scale:

0 = no requirements

1 = low requirements

3 = moderate requirements

5 = high requirements

These grades can then be quantified according to the respective national regulations (such as different time requirements concerning fire resistance).

Appropriate quality profiles are exemplified in Figs. 6.3 to 6.8, demonstrating a variety of quality demands for different types of technical subsystems as floors, walls, and roofs. Detailed information is provided for a floor in a multistory office building (Fig. 6.3), a floor in a multistory residential building (Fig. 6.4), and a floor over a cellar in a residential building (Fig. 6.5). An uninhabited attic floor in a residential building (Fig. 6.6), an outer wall in a multistory residential building (Fig. 6.7), and a roof over a factory building (Fig. 6.8) are also presented.

As the partial aspects of the indicated functional requirements (1 to 8) have a greater or lesser influence on the design of the building elements, they are subdivided further in the following examples. Variables of prime importance, which have to be taken into account, will be assessed for each individual case. The quality diagrams presented in this chapter are only models, as design fundamentals are not specified in detail. However, they illustrate the entire spectrum of possible quality demands.

Such quality profiles vary considerably in view of the numerous fields of application of plane building elements and that it is neither reasonable from the technical point of view nor justifiable from the economic point of view to design generally applicable building systems. On the other hand it is possible

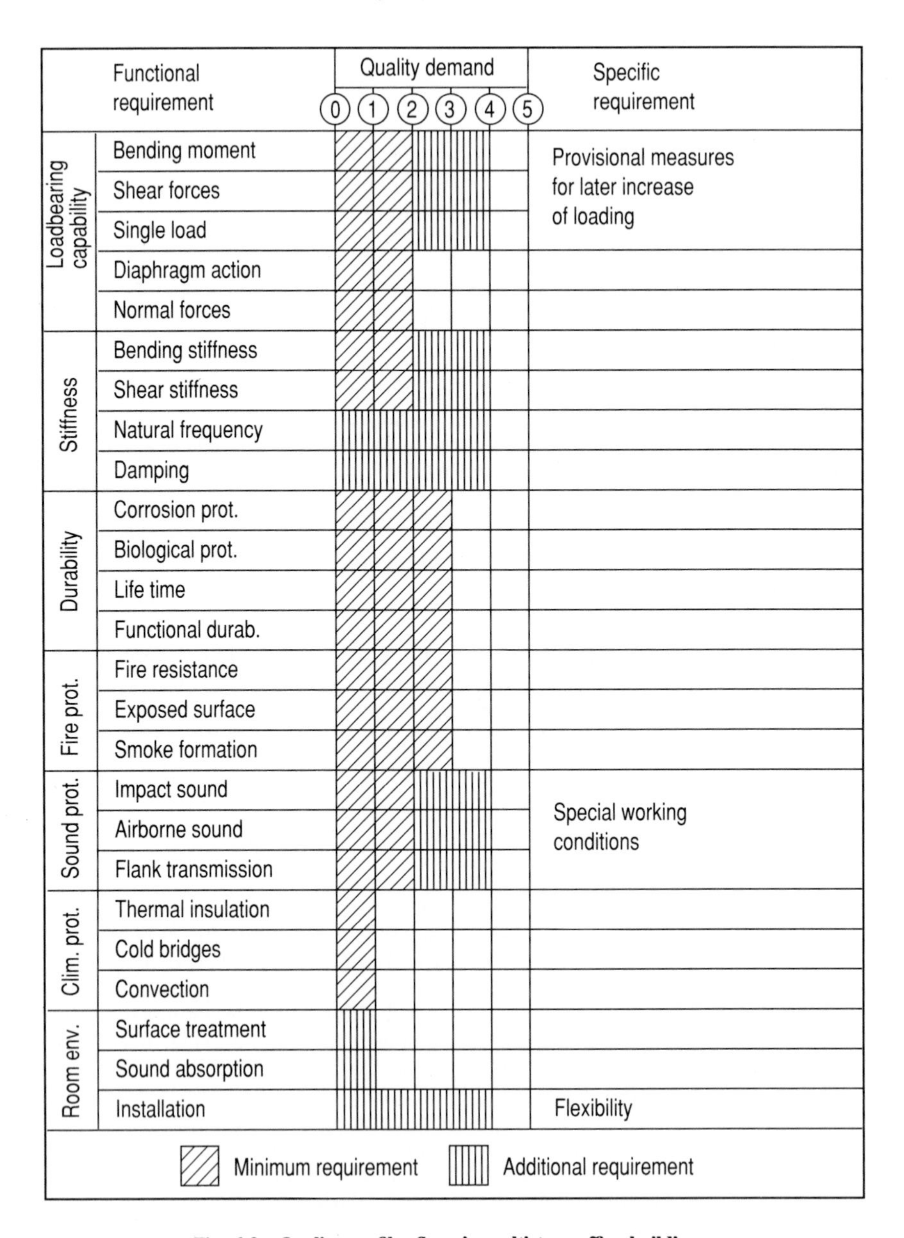

Fig. 6.3 Quality profile: floor in multistory office building.

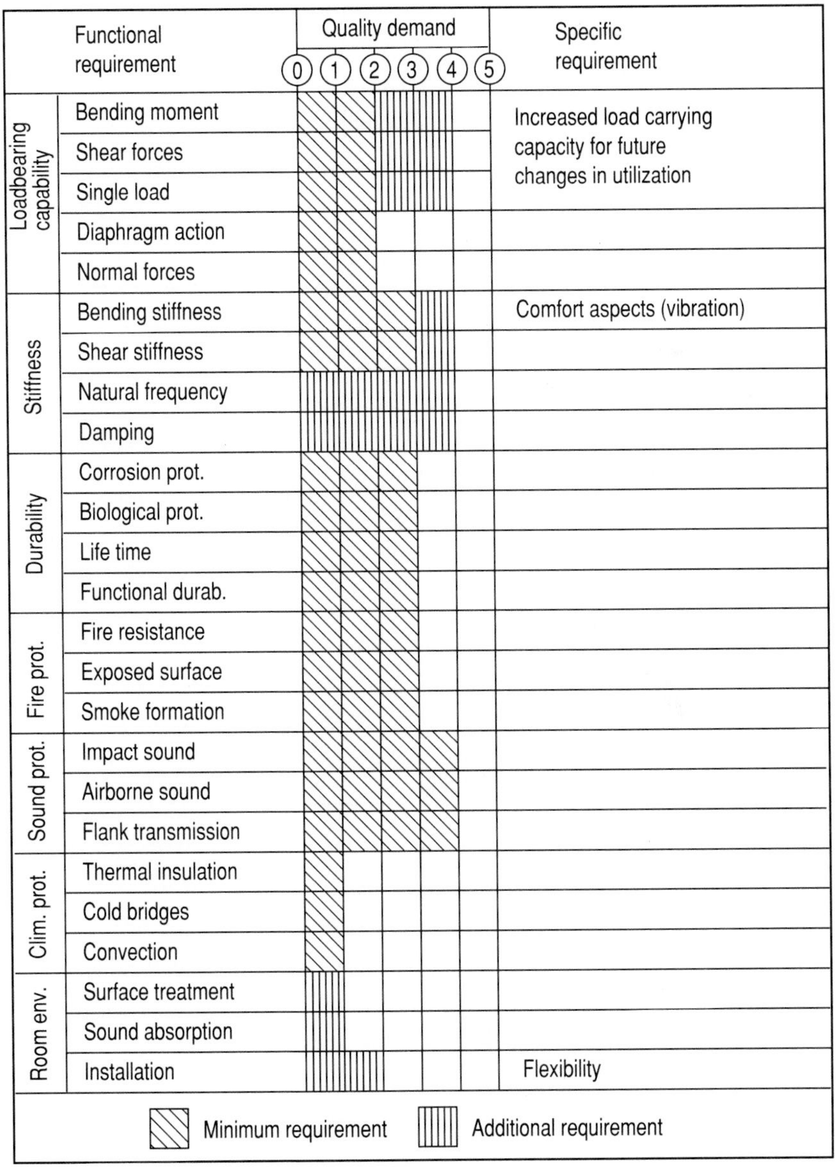

	Functional requirement	Quality demand 0–1	1–2	2–3	3–4	4–5	Specific requirement
Loadbearing capability	Bending moment	Min.	Min.	Add.	Add.		Increased load carrying capacity for future changes in utilization
	Shear forces	Min.	Min.	Add.	Add.		
	Single load	Min.	Min.	Add.	Add.		
	Diaphragm action	Min.	Min.				
	Normal forces	Min.	Min.				
Stiffness	Bending stiffness	Min.	Min.	Min.	Add.		Comfort aspects (vibration)
	Shear stiffness	Min.	Min.	Min.	Add.		
	Natural frequency	Add.	Add.	Add.	Add.		
	Damping	Add.	Add.	Add.	Add.		
Durability	Corrosion prot.	Min.	Min.	Min.			
	Biological prot.	Min.	Min.	Min.			
	Life time	Min.	Min.	Min.			
	Functional durab.	Min.	Min.	Min.			
Fire prot.	Fire resistance	Min.	Min.	Min.			
	Exposed surface	Min.	Min.	Min.			
	Smoke formation	Min.	Min.	Min.			
Sound prot.	Impact sound	Min.	Min.	Min.	Min.		
	Airborne sound	Min.	Min.	Min.	Min.		
	Flank transmission	Min.	Min.	Min.	Min.		
Clim. prot.	Thermal insulation	Min.					
	Cold bridges	Min.					
	Convection	Min.					
Room env.	Surface treatment	Add.					
	Sound absorption	Add.					
	Installation	Add.	Add.				Flexibility

Min. = Minimum requirement; Add. = Additional requirement

Fig. 6.4 Quality profile: floor in multistory residential building.

	Functional requirement	Quality demand 0–1	1–2	2–3	3–4	4–5	Specific requirement
Loadbearing capability	Bending moment	Min.	Min.	Add.	Add.	Add.	Increased load capacity
	Shear forces	Min.	Min.	Add.			
	Single load	Min.	Min.	Add.			
	Diaphragm action	Min.	Min.				
	Normal forces	Min.	Min.				
Stiffness	Bending stiffness	Min.	Min.	Add.	Add.		Stringent requirements concerning floor stiffness
	Shear stiffness	Min.	Min.	Add.	Add.		
	Natural frequency	Add.	Add.	Add.	Add.		
	Damping	Add.	Add.	Add.	Add.		
Durability	Corrosion prot.	Min.	Min.	Add.			Special protective measures (swimming pool)
	Biological prot.	Min.	Min.	Add.			
	Life time	Min.	Min.	Add.	Add.		
	Functional durab.	Min.	Min.	Add.	Add.		
Fire prot.	Fire resistance	Min.	Min.				
	Exposed surface	Min.	Min.				
	Smoke formation	Min.	Min.				
Sound prot.	Impact sound	Add.	Add.				Sound insulation (work room)
	Airborne sound	Add.	Add.	Add.			
	Flank transmission	Add.					
Clim. prot.	Thermal insulation	Add.	Add.				
	Cold bridges						
	Convection						
Room env.	Surface treatment	Add.	Add.	Add.			Extras
	Sound absorption	Add.	Add.	Add.			
	Installation	Add.					

Min. = Minimum requirement; Add. = Additional requirement

Fig. 6.5 Quality profile: floor over cellar in single-story residential building.

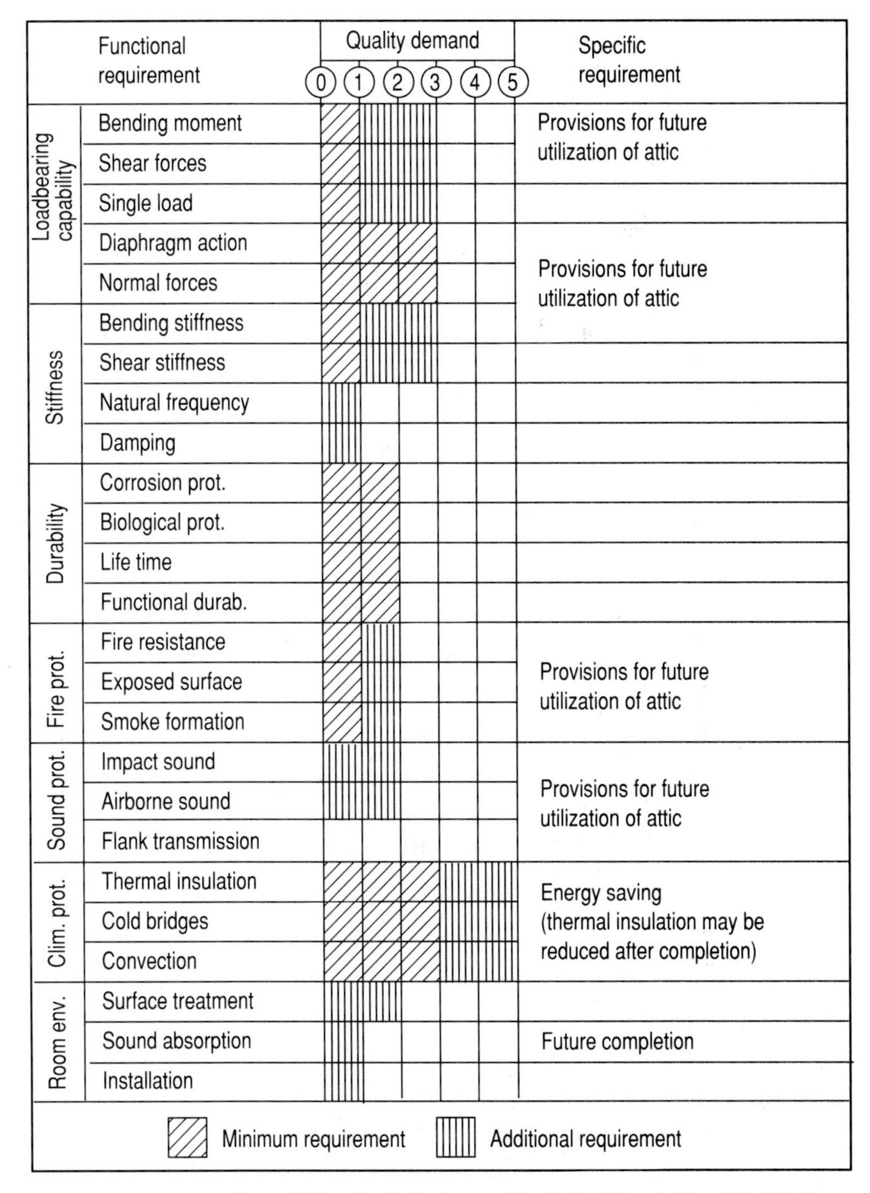

Fig. 6.6 Quality profile: attic floor in single-story residential building (attic uninhabited).

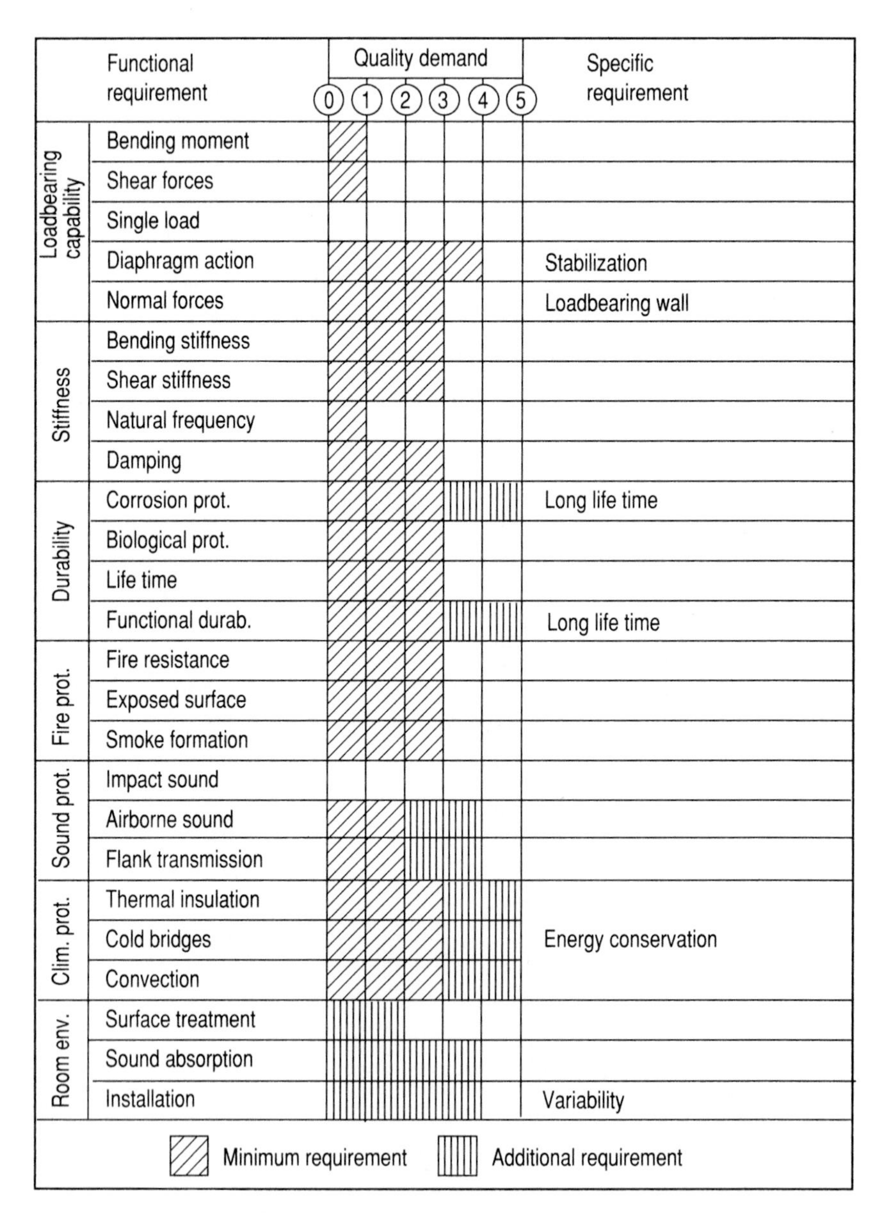

Fig. 6.7 Quality profile: outer wall in multistory building (housing).

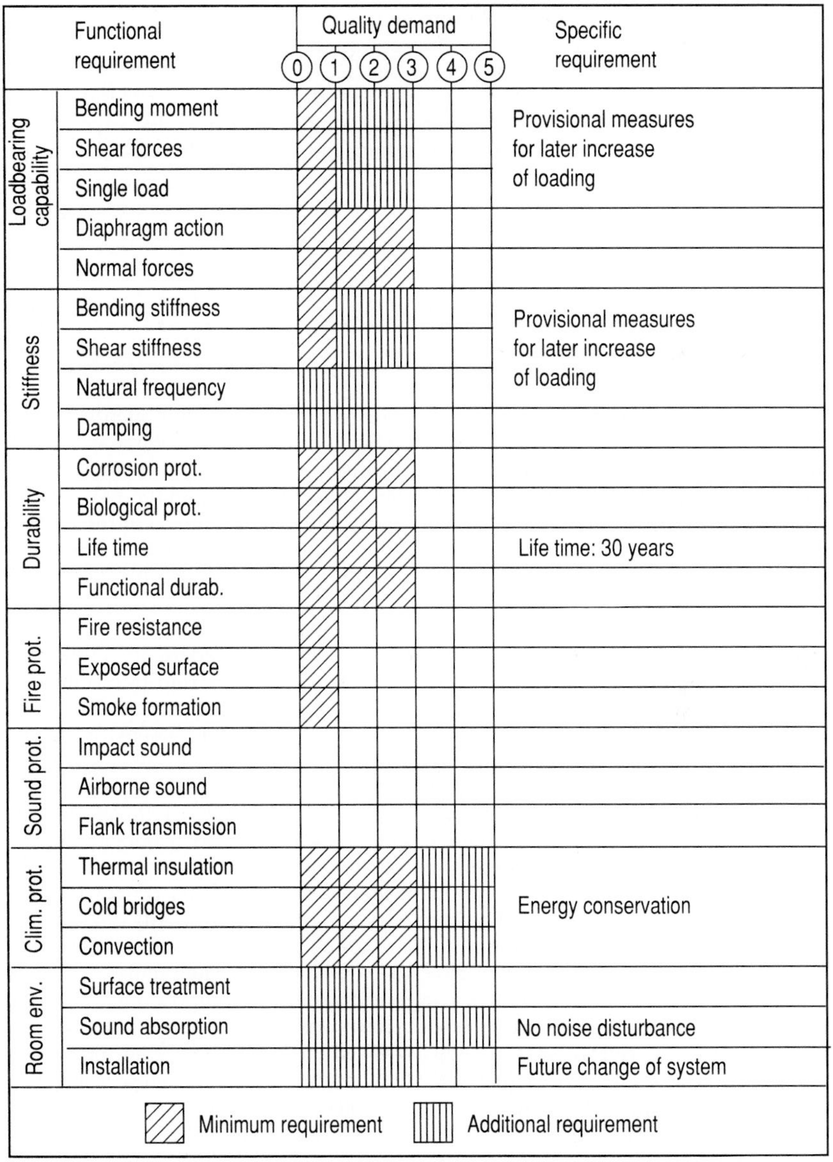

	Functional requirement	Quality demand 0–1	1–2	2–3	3–4	4–5	Specific requirement
Loadbearing capability	Bending moment	Minimum	Additional	Additional			Provisional measures for later increase of loading
	Shear forces	Minimum	Additional	Additional			
	Single load	Minimum	Additional	Additional			
	Diaphragm action	Minimum	Minimum	Minimum			
	Normal forces	Minimum	Minimum	Minimum			
Stiffness	Bending stiffness	Minimum	Additional	Additional			Provisional measures for later increase of loading
	Shear stiffness	Minimum	Additional	Additional			
	Natural frequency	Additional	Additional				
	Damping	Additional	Additional				
Durability	Corrosion prot.	Minimum	Minimum	Minimum			
	Biological prot.	Minimum	Minimum				
	Life time	Minimum	Minimum	Minimum			Life time: 30 years
	Functional durab.	Minimum	Minimum	Minimum			
Fire prot.	Fire resistance	Minimum					
	Exposed surface	Minimum					
	Smoke formation	Minimum					
Sound prot.	Impact sound						
	Airborne sound						
	Flank transmission						
Clim. prot.	Thermal insulation	Minimum	Minimum	Minimum	Additional	Additional	Energy conservation
	Cold bridges	Minimum	Minimum	Minimum	Additional	Additional	
	Convection	Minimum	Minimum	Minimum	Additional	Additional	
Room env.	Surface treatment	Additional	Additional	Additional			
	Sound absorption	Additional	Additional	Additional	Additional	Additional	No noise disturbance
	Installation	Additional	Additional	Additional			Future change of system

Minimum requirement (diagonal hatching) — Additional requirement (vertical hatching)

Fig. 6.8 Quality profile: roof over factory (light industry).

to develop—on the basis of the lowest common denominator—a basic system which can be improved to satisfy the different quality demands.

Furthermore, the quality profile is a checklist which forces the systematic analysis of partial aspects and the assessment of effects of product variations on other partial aspects. After quantification of the quality demands, the expenditure for different solutions can be estimated, taking into account the materials to be used and the design.

6.3 BASIC PRODUCTS AND COMPOSITE MATERIALS

The main load-bearing function with reference to functional requirements 1 and 2 is to be provided by cold-formed sections (Fig. 6.9), often in combination with adequate formed sheet products as, for example, trapezoidal sheeting, or by sandwich panels (Fig. 6.10). Requirements concerning durability (3) can be satisfied by zinc protection and, if necessary, by additional coatings in the case of special environmental conditions. In order to satisfy other functional requirements, such as fire protection, sound insulation, and climatic protection, supplementary materials are necessary.

According to Fig. 6.10, only partial requirements can be satisfied by the materials marketed at present. However, by combining adequate composite materials, optimum general solutions can be found. Besides thin fiber concrete board, mineral and wood-based materials are mostly used. At present, however, there is only limited reliable information on the long-term behavior of the physical properties and the effects of stress history for such materials.

Table 6.1 lists guide values for the mechanical properties in view of a possible combination with steel components. The properties of the previously mentioned materials concerning functional requirements 3 to 6 depend on the type of product, and are partly defined in official approvals.

As to product design, it should be considered that the given properties only indicate the behavior of the building elements, but do not permit any quantitative statement on how the quality demands are satisfied. In many cases the quality of the respective building element will have to be verified experimentally.

The connection between the steel components and the composite material can be rigid or flexible, achieved by mechanical fasteners (self-tapping screws, self-drilling screws, rivets, bolts, or powder-actuated fasteners), or, in the case of prefabrication, by adequate adhesives (such as on a polyurethane basis) and perhaps in combination with screws or bolts. The composite action, the load-bearing behavior, and the flexibility of the connections have to be determined experimentally (see Chapter 5).

The design basis is the general theory of composite elements where the geometrical parameters of the composite section are referred to the modulus of elasticity of the steel section. The time- and stress-dependent physical properties of the composite material have to be taken into account. Due to the comparatively low moduli of elasticity, it seems necessary to develop special shapes for composite systems. As the thicknesses of the mentioned board materials are comparatively small, the effective width of a cross section in compression is limited accordingly [$b_e = 200$ to 400 mm (8 to 16 in.)]. The composite action can

Structural elements, suitable for compression or tension forces

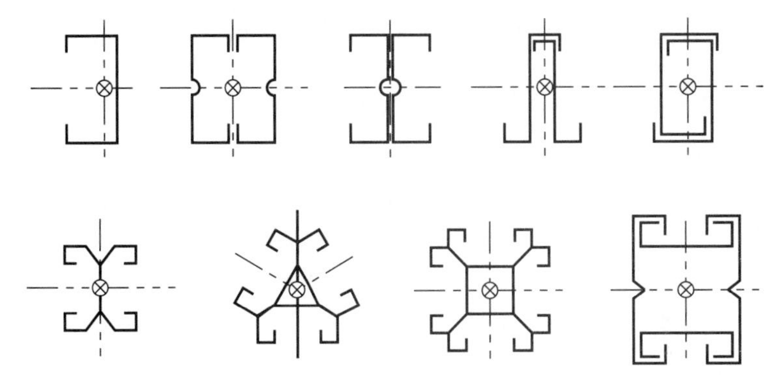

Structural elements, suitable for bending moments

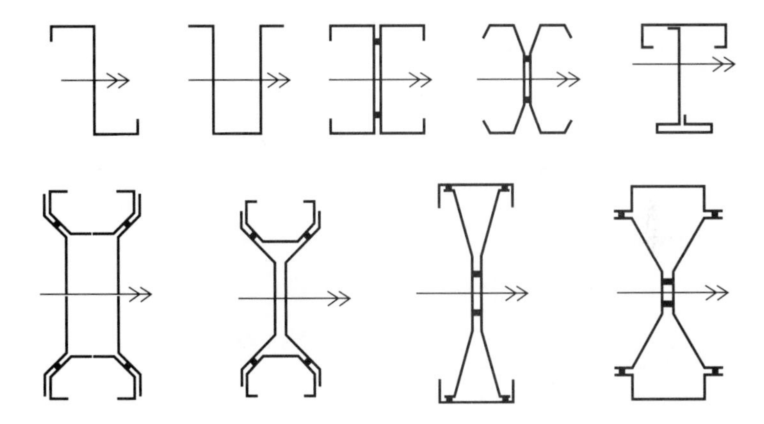

Space-covering elements, suitable for combined actions

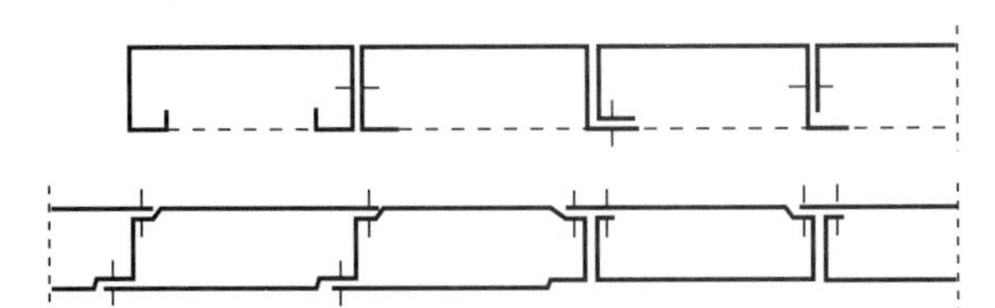

Fig. 6.9 Building elements based on cold-formed sections.

therefore be reasonably defined as:

1. Reinforcement of thin-walled sheet steel susceptible to buckling
2. Linear composite action for cold-formed sections and beams
3. Two-way composite action for sheeting and panels

The effect of a plane composite structure (2) can be illustrated by means of the equivalent value of the steel thickness t_{si} for different plate materials and thicknesses (Table 6.1). Here, in particular for plywood boards, the value of t_{si} is on the order of magnitude of the actual sheet thickness [$t = 1.0$ mm (0.04 in.)].

Building materials:

- ■ Suitable for high requirements
- □ Suitable for moderate requirements
- (blank) Suitable for low requirements
- — Unsuitable

	Building materials	Functional requirements						
		1	2	3	4	5	6	7
		Load bearing capacity	Stiffness	Durability	Fire protection	Sound protection	Climatical protection	Room environment
1	Unstiffened sheet panels (galv.)	□	□	□	—	—	—	—
2	AS 1. galv. • coated	□	□	■	—	—	—	—
3	Stiffened sheet panels (galv.)	■	■	□	—	—	—	—
4	AS 3. galv. • coated	■	■	■	—	—	—	—
5	20-mm fiber reinforced concrete	□	□	■			—	
6	13-mm gypsum board			□	□		—	□
7	26-mm gypsum board	□	□	□	■	□	—	■
8	13-mm plywood	□	□	□			—	□
9	Fiber board					—	—	■
10	Chipboard	—	—		—			□
11	Fiber board (mineral binding)	□	□	□	■		—	■
12	Mineralwool	—	—	□	■	■	■	—
13	Plastic foam			□		□	■	—

Fig. 6.10 Protective qualities of composite materials.

The effect of other materials must not be underestimated if compressed zones susceptible to buckling are stiffened (1), as this increases the effective width of the steel section. These ratios are shown in Fig. 6.11 for a C section in bending (types 1 to 4).

Case (2) highlights the fact that the board materials act as the compression flange of the composite structure and that the load is transmitted transversely. With regard to this factor and to the limited effective width of the plate material, the spacing of linear load-bearing members should not exceed about 400 mm (16 in.) for wooden boards. The linear composite action (2) is therefore advantageous if cold-formed sections are used, as illustrated for an inverted C section with 12-mm (0.5-in.) plywood in the compressed zone (Fig. 6.11, type 5).

The following fundamental conclusions can be drawn from this comparison:

Using wooden or gypsum boards with two-way action, the stiffness of a thin-walled steel section susceptible to buckling can be raised to a value corresponding at least to the stiffness of an unbuckled section.

In the case of linear composite action, with regard to the load-bearing capacity of the composite material and the appropriate profile geometry, the stiffness can be increased considerably.

Table 6.1 Mechanical properties of composite materials and equivalent values of plate thickness referred to sheet steel, $t_{si} = d/n$

Material, product type	Plate thickness d, mm	Weight per unit area, kg/m^3	Characteristic strength value, N/mm^3	Young's modulus (compression), N/mm^3	$n = \frac{E_S}{E_V}$	Equivalent plate thickness t_{s1}, mm
1 Steel sheet	1.0	8	$\beta_s = 280–350$	210,000	1.0	1.0
2 Sandwich-type plasterboard	12.5	13	$\beta_B = 7$	2,000	105	0.12
3 Sandwich-type plasterboard	18.0	18	$\beta_B = 5$	2,000	105	0.17
4 Fiber concrete–steel fibers	20	50	$\beta_D = 600$	21,000	10	2
5 Asbestos cement sheeting	4	7	$\beta_B = 26$	18,000	12	0.33
6 Plywood board AW 100	13	8	$\beta_B = 20$	7,000	30	0.43
7 Plywood board AW 100	22	15	$\beta_B = 20$	7,000	30	0.73
8 Wood fiber board (hard)	10	8	$\beta_B = 35$	2,000	105	0.1
9 Particleboard	32	20	$\beta_B = 12$	2,000	105	0.3
10 Chipboard (mineral binder)	12	15	$\beta_D = 15$	3,000	70	0.17
11 Mineral fiberboard (mineral binder)	12	11	$\beta_D = 11$	3,500	60	0.2

The same applies to the load-bearing capacity, provided that the composite action is secure until the limit state is reached (design load).

With regard to the fact that in the serviceability state the stresses in the composite section are relatively low and the stiffness of lightweight floors is of particular importance, it is efficient to actively utilize the increased stiffness resulting from composite action.

6.4 EXAMPLES OF BUILT-UP STRUCTURAL ELEMENTS

With reference to adequate functional requirements and quality demands according to Section 6.2, three different building elements are discussed. Their common feature is that the load-bearing capacity and the stiffness are provided by the composite action of steel sheet sections and (nonmetallic) board materials. The composite action is provided by adhesive-bonded joints in combination with nails and drilling screws.

	Section	Effective Section	Equivalent flange thickness t_{Si} (mm)	Relative moment of inertia I/I_{St}
1	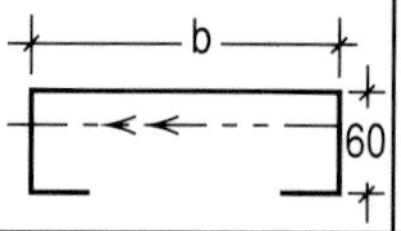 b; 60	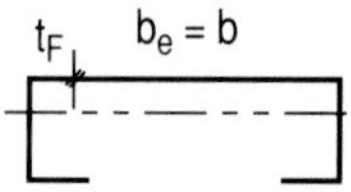 t_F, $b_e = b$	0,7	2,3
2		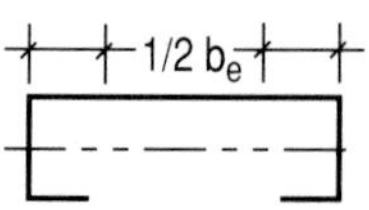 1/2 b_e	0,7	1,0
3	12-mm plywood 	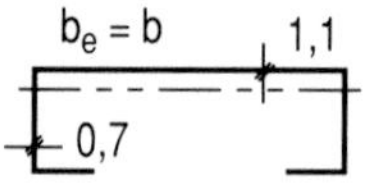 $b_e = b$; 1,1; 0,7	0,7 + 0,4 = 1,1	2,6
4	13-mm gypsumboard 	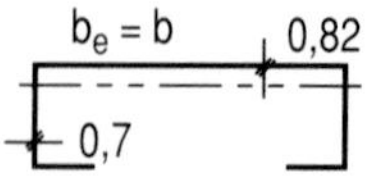 $b_e = b$; 0,82; 0,7	0,7 + 0,12 = 0,82	2,4
5	12-mm plywood 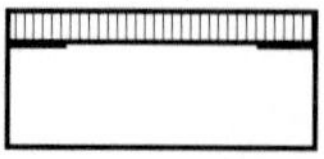	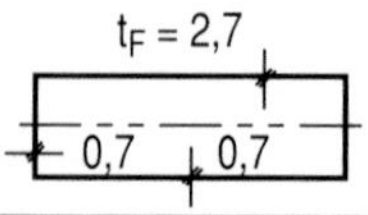 $t_F = 2,7$; 0,7; 0,7	0,7 + 2,0 = 2,7	4,7

Fig. 6.11 Comparative values of moments of inertia of composite sections.

Table 6.2 Example of floor structure in multistory office building*

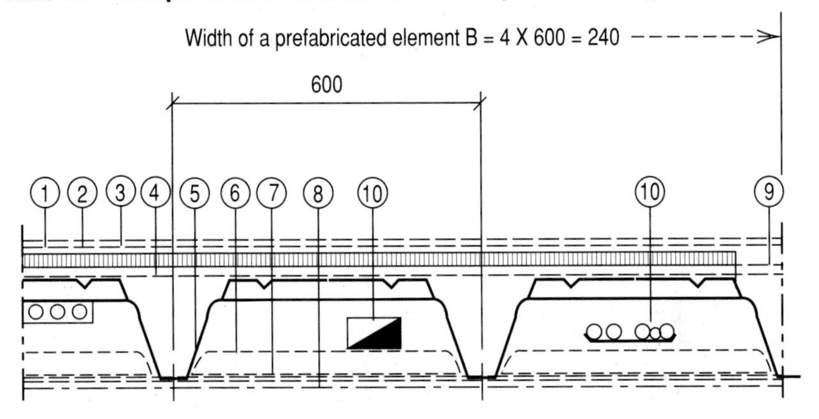

Components: (Shop fabrication (2)-(7))

(1) Carpet floor
(2) Hard particle board
(3) 13-mm mineral fiber board
(4) 13-mm plywood board connected with (5) by self-drilling screws and adhesive
(5) Floor section of steel sheeting
(6) Mineral wool board
(7) Transverse stiffeners
(8) 13-mm gypsum board
(9) Compensating element on support angles (site assembly)
(10) Installations

Functional requirements	Material provisions	Functional behavior
Loadbearing capacity	Cold-formed channels (5): $t = 1.0$ mm	Allowable bending moment and shear force provided by the sections only. Composite action together with plywood board taking advantage of increased stiffness (reduced deflection, increased natural frequency) and provision of in-plane shear capacity. Utilization of shear flexibility in (3) for increased damping.
Stiffness	Cold-formed channels (5) in composite action with plywood board (4). Damping provided by mineral fiber board (3) connected to hard particle board.	
Durability	Zinc protection (Z 275).	Sufficient under normal in-door humidity conditions.

* Functional requirements and quality demands according to Fig. 6.3. Span $L = 5.0$ m; load $q = 3$ kN/m^2.

Fire protection is, for all types, provided primarily by gypsum boards with thicknesses depending on the required resistance duration, whereas climatic protection is achieved mainly by mineral wool products. Sound insulation, where required, can be achieved by board materials on the one hand and by suitable separation of load-bearing elements on the other hand.

Measures to fulfill requirements for floor, wall, and roof structures are listed in Tables 6.2 to 6.4. It should be taken into account that these design examples, too, are only models, and that different solutions are possible. The floor structure of Table 6.2 refers to the requirements of Fig. 6.3, the wall structure in Table 6.3 to Fig. 6.7, and the roof structure in Table 6.4 to Fig. 6.8.

It is obvious that the various composite materials are able to fulfill different functional requirements and—on the other hand—that different functional demands have partly contradictory consequences for the material choice and the structural design, which requires compromise solutions.

The consequences for product development can be summarized as follows.

Stage 1. Preliminary study of functional requirements

1. Fields of application (such as office buildings, residential buildings)
2. Actual life loads (standards and regulations)
3. Actual span ranges (assessment)
4. Supply and removal facilities in floor zone (assessment)
5. Requirements concerning fire protection, sound insulation, and weather-proofing (standards)

Table 6.2 Example of floor structure in multistory office building (*Continued*)

Functional requirements	Material provisions	Functional behavior
Fire protection	*Upside*: hard particle board (2). *Downside*: double layer of gypsum board (8).	Fire resistance 90 min. (to be verified by fire test).
Sound insulation	Multilayer structure combined with "floating" facings. Additional textile cover (1) and mineral wool products (6).	Airborne sound transmission prevented by intermediate layers of flexible material. Structural sound transmission prevented by interrupting sound bridges and flanking transmission (to be verified by test).
Weatherproofing	Mineral wool products (as stated above).	Low requirements
Room environment	*Upside*: textile cover on hard particle board. *Downside*: gypsum board.	Space for installations provided between floor structure and ceiling.

* Functional requirements and quality demands according to Fig. 6.3. Span $L = 5.0$ m; load $q = 3$ kN/m^2.

Table 6.3 Example of load-bearing wall structure for multistory office building (gable wall)*

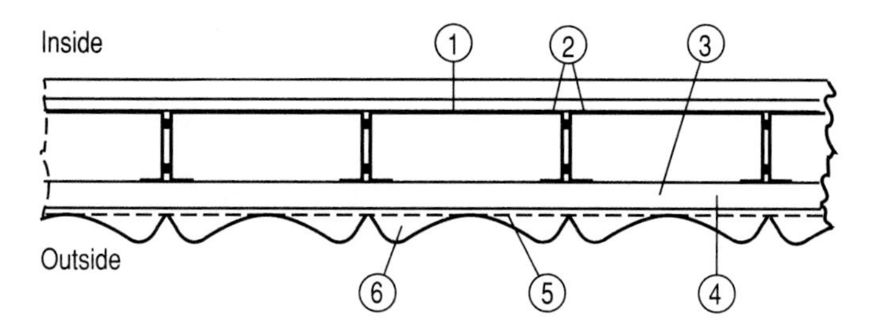

1. Gypsum board (glued or screwed to the section)
2. Cold-formed C-profiles (300 X 100 X 25 t = 1.0mm connected by spot welding)
3. Mineral wool (100 + 50mm)
4. Horizontal rail 50 X 50 or cold-formed Z-profiles 25 X 50 X 25
5. Wind protection
6. Decorative sheeting

Functional requirements	Material provisions	Functional behavior
Loadbearing capacity	Cold-formed channels (2) and gypsum board (1) in composite action.	Composite action of (1) and (2) in order to prevent local buckling in the wide flange. Loadbearing functions: transverse load (q) due to wind pressure and suction; vertical load (N) due to gravity loads from the floor structure; horizontal in-plane load (H) from wind load. Sufficient in-plane stiffeners against sway provided.
Stiffness	Composite action provided by connections (adhesive and screws) between (1) and (2) and spot welds between the channels.	
Durability	Zinc protection on channels (2). Aluminum sheeting or plastic coated steel sheeting outside.	Acceptable for normal climatic conditions inside and outside.
Fire protection	Double layer of gypsum board inside (1).	Fire resistance 90 min.
Sound insulation	Double layer of mineral wool (5) and double layer of gypsum board inside (1).	Airborne sound from outside prevented by mineral wool. Sound bridges causing structural sound transmission from inside to be avoided.
Weatherproofing	Double layer of mineral wool (3) and layer of wind protection (5) in order to prevent convection.	Sufficient thermal insulation, provided that cold bridges can be reduced, e.g. by wooden horizontal rails (4) instead of cold-formed Z-sections.
Room environment	Gypsum board (1) inside.	Moisture-balancing material.

* Functional requirements and quality demands according to Fig. 6.7. Span $L = 3.0$ m; vertical load $N_{max} = 50$ kN/m; in-plane load $H_{max} = 10$ kN/m; transverse load $q = 0.6$ kN/m^2.

Table 6.4 Example of roof structure (factory)*

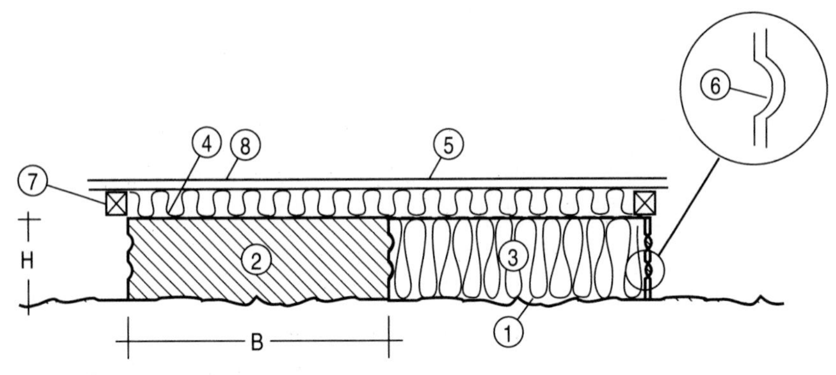

(1) Roof panel B 400 - 600 mm, H 70 - 160 mm, t 0.6-1.2 mm
(2) On-line produced polyurethane foam
(3) Mineral wool, appli ed at the workshop or on site
(4) Protection against rain, applied at the workshop
(5) Additional insulation layer of mineral wool or fiber
(6) Vapor barrier by sealing if necessary
(7) Distance rail of wood or other coldbridge insulation (e.g. hard mineral fiber board)
(8) Outer skin of trapezoidal sheeting or standing seam type

Functional requirements	Material provisions	Functional behavior
Loadbearing capacity	Cold-formed C-shaped panels (1) ($t = 1.0$ mm, $H = 160$ mm).	Allowable bending moment and shear force provided by the sections. Composite action with polyurethane foam in order to prevent buckling of the wide flange and to reduce initial deflections of the flange. Stiffening effects on the webs can be achieved.
Stiffness	Increased stiffness of the wide flange by on-line produced polyurethane foam.	
Durability	Zinc-protected steel sheeting inside. Zinc-protected and plastic-coated steel sheeting outside (8).	Zinc protection inside sufficient under normal humidity and environmental conditions; otherwise additional plastic coating.
Fire protection	No special measures necessary.	
Sound insulation		If necessary, with respect to inside sound emission, the wide flange can be perforated.
Weatherproofing	Foam (2) or mineral wool (3) inside the panels. Additional insulation on the top (5). Air-tightening (vapor barrier) in wet corrugations (6).	Distance rails between outer skin and structure of wood, or other cold-bridge insulation.
Room environment		Plastic coating for color effects and/or perforation for acoustic effects.

* Functional requirements and quality demands according to Fig. 6.7. Span $L = 8.0$ m; load $q = q + p = 0.3 + 1.0 = 1.3$ kN/m^2.

Stage 2. Definition of quality demands

1. Specification of actual functional requirements (application-oriented)
2. Definition of minimum requirements (regulations)
3. Consideration of application-oriented quality improvements
4. Preparation of a quality profile

Stage 3. Preliminary design of a floor unit

1. Consideration of material choice with respect to functional demands
2. Consideration of the composite action of different materials
3. Preliminary studies of manufacturing process
4. Rough estimates for the determination of load-bearing capacity and stiffness
5. Preparation of a property profile (analogous to quality profile)
6. Comparison of property and quality profiles and study of the weak points of the chosen systems

Stage 4. Product design

1. Choice of materials and components for the basic elements, satisfying the minimum requirements
2. Static calculation for the ultimate load states—serviceability and load-bearing capacity
3. Specification of additional measures to satisfy higher functional demands
4. Definition of the manufacturing process and decisions about prefabrication and in situ workmanship
5. Fabrication of prototypes for experimental verification of required qualities
6. Product modification on the basis of the test results

Stage 5. Product approval by the supervising authority, including production control

6.5 CONDENSED REFERENCES / BIBLIOGRAPHY

Baehre 1978, *Development Characteristics of Thin-Walled Building Technics—Section Stiffening, Components, Composite Action*

Baehre 1982, *Space Covering Building Elements*

Balasz 1980, *Stressed Skin Action in Composite Panels Comprising Steel Sheeting and Boards*

Balasz 1985, *Tests on Box Units of Steel Sheeting in Partial Structure Interaction with Plaster Board*

Council on Tall Buildings, Group SC 1980, *Tall Building Systems and Concepts*

ECCS 1984, *Lightweight Steel Based Floor Systems for Multi-Storey Buildings*

König 1981, *The Composite Beam Action of Cold-Formed Sections and Boards*

7

Stainless-Steel and Corrosion-Resisting Steel Sections

The objective of this chapter is to summarize the state of the art of the use of sections cold-formed from stainless and corrosion-resisting steels in tall buildings. The architectural use of stainless-steel sections for their pleasing appearance, superior corrosion resistance, and ease of maintenance is illustrated in Figs. 7.1 to 7.4 by means of typical examples. Load-carrying structural members cold-formed from stainless steels were previously designed in accordance with the *Specification for the Design of Cold-Formed Stainless Steel Structural Members,* published by American Iron and Steel Institute (AISI, 1974a). Prior to 1990 it was the only specification for the design of such members and was contained as Part I in the 1974 edition of the *Stainless Steel Cold-Formed Structural Design Manual* (AISI, 1974b). The design manual also contains a commentary in Part II, supplementary information in Part III, design examples in Part IV, and tables and charts, useful for determining the safe load-carrying capacities and deflections of sections, in Part V.

In 1990 the American Society of Civil Engineers (ASCE) issued a new *Specification for the Design of Cold-Formed Stainless Steel Structural Members* (1991), which was developed at the University of Missouri-Rolla under the sponsorship of ASCE. Financial support was provided by Chromium Centre, Nickel Development Institute, and the Specialty Steel Industry in the United States. This new ASCE standard specification includes both the load and resistance factor design (LRFD) method and the allowable stress design (ASD) method. In the LRFD method, separate load and resistance factors are applied to specific loads and nominal resistance to ensure that the probability of reaching a limit state is acceptably small. These factors reflect the uncertainties of analysis, design, loading, material properties, and fabrication.

The types of stainless steels for which that document is valid, as well as the typical stress-strain behavior of those materials, are briefly discussed in this chapter. Due to the differences in the mechanical properties of the stainless steels compared with the carbon and low-alloy steels, different design rules have been adopted in certain cases for the stainless steels. These differences are highlighted amidst a general, summarized discussion of the design provisions for stainless-steel members. Finally, the development of certain corrosion-resisting steels, as well as the need for establishing design criteria for such steels, is discussed.

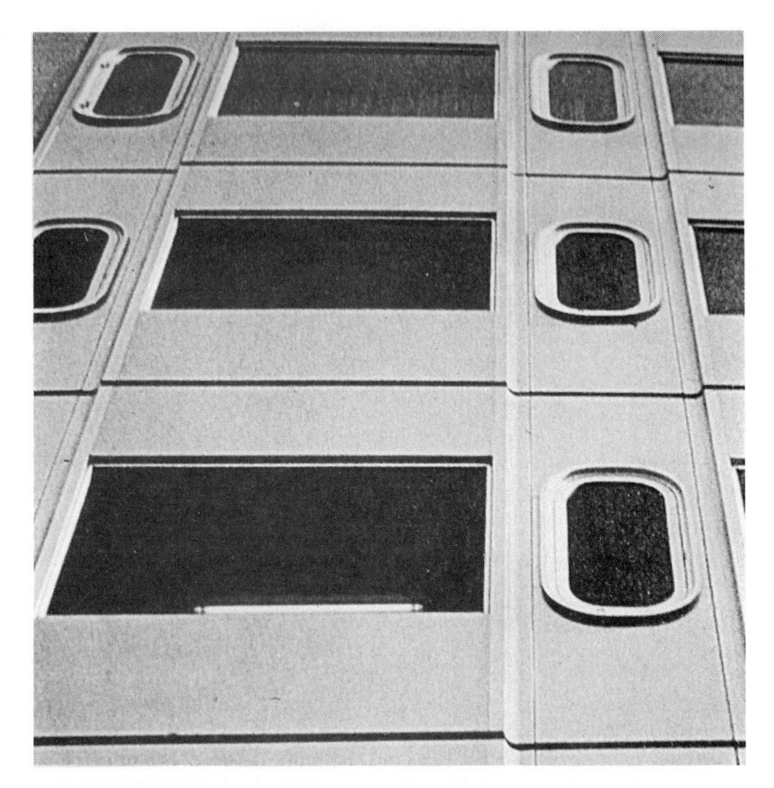

Fig. 7.1 Cladding; type AISI 304 (Canada).

Fig. 7.2 Roofing; type AISI 304 (Italy).

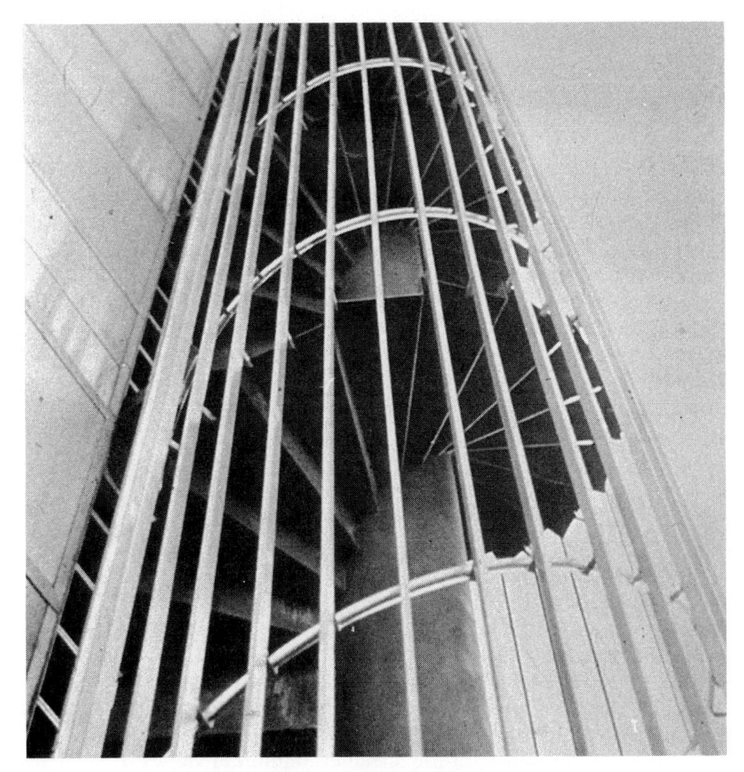

Fig. 7.3 Tower; type AISI 304L (Canada).

Fig. 7.4 Church building; type AISI 304 (Japan).

7.1 ASCE STANDARD FOR STAINLESS STEEL

The ASCE (1991) standard ANSI/ASCE-8-90 is based on research conducted at Cornell University (Johnson, 1966a; Johnson and Winter, 1966b; Wang, 1969; Errera et al., 1970), the University of Missouri-Rolla (van der Merwe, 1987; Lin, 1989), and Rand Afrikaans University (van den Berg, 1988), as well as on the accumulated experience in the design of cold-formed carbon steel structural members (AISI, 1986). It is primarily intended for the design of cold-formed stainless-steel structural members used for load-carrying purposes in buildings and other statically loaded structures. It may also be used for the design of structures that are subjected to dynamic loads, provided appropriate allowances are made for such effects. This ASCE standard supersedes the 1974 edition of the AISI specification for the design of stainless-steel structural members.

7.2 MATERIAL

The new ASCE specification is applicable for AISI types 201, 301, 304, and 316 (annealed, $^1/_{16}$, $^1/_4$, and $^1/_2$ hard) austenitic stainless steels and types 409, 430, and 439 (annealed) ferritic stainless steels. Other stainless steels may be used provided that they conform to the chemical and mechanical requirements of one of the material specifications listed in the ASCE standard or other published specification which establishes properties and suitability.

A comparison of the ranges of the specified minimum values of yield strength, ultimate strength, ratios of ultimate strength to yield strength, and percent elongation between the austenitic stainless steels and the carbon and low-alloy steels listed in the respective AISI design specifications, is made in Table 7.1. A detailed comparison of the 15 types of carbon and low-alloy steels, available in 49 different strength levels, has been made by Yu (1991). Even though the ratios of the specified minimum values of ultimate strength to yield strength for the carbon and low-alloy steels are in the range of 1.17 to 2.22 and the specified minimum percentages of elongation range from 12 to 27, the specification for carbon and low-alloy steels (AISI, 1986) allows that the ratio F_u/F_y be a minimum of 1.08 and the elongation in 50-mm (2-in.) gage length be a minimum of 10% for steels other than those listed in that specification. A similar relaxation on ductility requirements exists in the stainless steel design specification.

Table 7.1 Ranges of specified minimum properties

Property	Carbon and low-alloy steels	Austenitic stainless steels
Yield strength (tension) F_y	25–70 ksi (172–483 MPa)	30–110 ksi (205–760 MPa)
Ultimate strength F_u	42–85 ksi (290–586 MPa)	75–150 ksi (515–1035 MPa)
F_u/F_y	1.17–2.22	1.36–2.50
Elongation	12–27%	7–40%

7.3 MECHANICAL PROPERTIES OF LISTED STAINLESS STEELS

The stress-strain behavior of the listed austenitic stainless steels has been studied by Johnson (1966a), Johnson and Kelson (1969), and Wang (1969). Figure 7.5 compares the stress-strain curves of annealed and strain-flattened and half-hard austenitic stainless steels. For a carbon or low-alloy steel, a single stress-strain curve of the sharp yielding type (for virgin material) is assumed to be valid for tension and compression, irrespective of the direction of rolling of the flat product. In contrast to this, an austenitic stainless steel has four distinct normal stress-strain curves at each strength level. These are identified as follows: longitudinal tension (LT), longitudinal compression (LC), transverse tension (TT), and transverse compression (TC). The term longitudinal refers to the direction parallel to the direction of rolling and the term transverse implies the direction perpendicular to the direction of rolling.

The gradual yielding nature of the stress-strain behavior is quite evident from

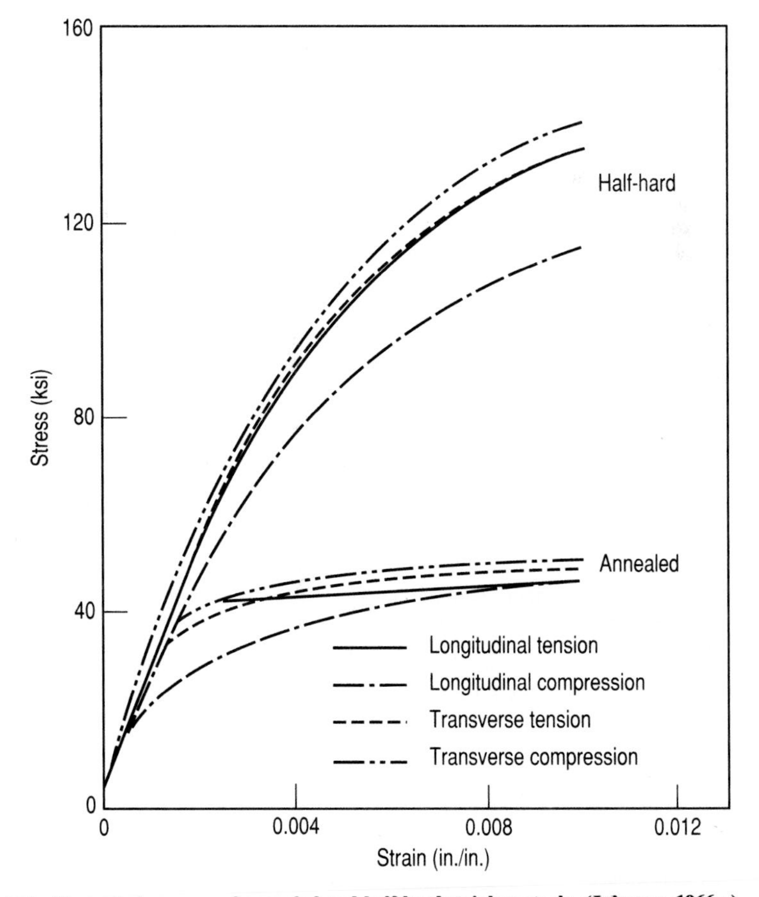

Fig. 7.5 Stress-strain curves of annealed and half-hard stainless steels. (Johnson, 1966a.)

Fig. 7.5. Also evident, although not indicated, are the relatively low proportional limits (0.01% offset strength) in relation to the yield strengths (0.2% offset strength), especially for longitudinal compression. Table 7.2 lists the values of yield strength for tension, compression, and shear (ASCE, 1991). The ratios of proportional limit to yield strength are given in Table 7.3 (ASCE, 1991).

The various values of the initial moduli of elasticity E_0 and the initial shear moduli G_0 are given in Table 7.4 (ASCE, 1991).

Due to the differences in stress-strain behavior between the austenitic stainless steels and the carbon and low-alloy steels, the various design specifications, covering the design of carbon and low-alloy steel structural members, do not apply to the design of stainless-steel structural members.

Table 7.2 Specified yield strengths F_y of stainless steels in ksi (MPa)

	Types 201, 301, 304, 316						
Type of stress	Annealed		$^1/_{16}$ hard	$^1/_4$ hard	$^1/_2$ hard	Type 409	Types 430, 439
Longitudinal tension	30 (206.9)	45* (310.3)	40† 45 (275.8)	75 (517.1)	110 (758.5)	30 (206.9)	40‡ (275.8)
Transverse tension	30	45*	40† 45	75	110	35‡ (241.3)	45‡ (310.3)
Transverse compression	30	45*	40† 45	90	120	35‡	45‡
Longitudinal compression	28 (193.1)	41* (282.7)	36† 41 (248.2)	50 (344.8)	65 (448.2)	30	40‡
Shear yield strength F_{yv}	17 (117.2)	25* (172.4)	23† 25 (158.6)	42 (289.6)	56 (386.1)	19 (131)	24 (165.5)

* For type 201-2 (class 2).
† Flat bars, for type 201 only.
‡ Adjusted yield strengths; ASTM-specified yield strength is 30 ksi (206.9 MPa) for types 409, 430, 439.

Table 7.3 Ratios of effective proportional limit to yield strength F_{pr}/F_y

	Types 201, 301, 304, 316				
Type of stress	Annealed and $^1/_{16}$ hard	$^1/_4$ hard	$^1/_2$ hard	Type 409	Types 430, 439
Longitudinal tension	0.67	0.50	0.45	0.76	0.70
Transverse tension	0.57	0.55	0.60	0.83	0.81
Transverse compression	0.66	0.50	0.50	0.83	0.82
Longitudinal compression	0.46	0.50	0.49	0.73	0.62

7.4 DESIGN PROCEDURE

Cold-formed stainless-steel design is similar to cold-formed carbon steel design. However, because the mechanical properties of austenitic stainless steels are more complex than those of carbon steel, the design procedures are occasionally somewhat more involved. The ASCE standard contains the load and resistance factor design in Sections 1 through 6; the allowable stress design method is included in Appendix E (ASCE, 1991).

7.5 ANISOTROPY

During design and fabrication allowance must be made for the orientation of the member element or member with respect to the direction of rolling of the flat plate, sheet, or strip material. The lengths of members, which were cold-formed such that the properties transverse to rolling may be used, are limited to the available widths of plates and sheets. For this reason, the properties applicable to the longitudinal direction are most frequently used.

7.6 DEFLECTION DETERMINATION

A reduced modulus E_r is used when deflections of stainless-steel members are computed. E_r is evaluated as

$$E_r = E_{st} + E_{sc}, \tag{7.1}$$

where E_{st} and E_{sc} are the secant moduli corresponding to the stress in the tension and compression flanges, respectively.

Table 7.4 Initial moduli of elasticity E_0 and shear moduli G_0

	Types 201, 301, 304, 316			Types 409, 430, 439	
Modulus	Annealed and $^1/_{16}$ hard, longitudinal and transverse tension and compression	$^1/_4$ and $^1/_2$ hard, longitudinal tension and compression	$^1/_4$ and $^1/_2$ hard, transverse tension and compression	Longitudinal tension and compression	Transverse tension and compression
Initial modulus of elasticity E_0, ksi × 10^3 (MPa × 10^3)	28.0 (193.1)	27.0 (186.2)	28.0 (193.1)	27.0 (186.2)	29.0 (200)
Initial shear modulus G_0, ksi × 10^3 (MPa × 10^3)	10.8 (74.5)	10.5 (72.4)	10.8 (74.5)	10.5 (72.4)	11.2 (77.2)

7.7 OTHER CORROSION-RESISTING STEELS

Stainless steels are relatively highly priced alloy steels, especially the austenitic stainless steels that contain molybdenum in addition to substantial quantities of nickel and chromium. Even so, the stainless steels are competitive in many applications (Melvill et al., 1980), particularly when the true cost is computed, which is initial material cost plus coating, inspection, and maintenance costs.

A number of specialty-steel producing companies in the United States, Europe, Japan, and South Africa have attempted to develop a low-priced weldable chromium-containing steel, having a corrosion resistance ability between that of the established grades of stainless steel and that of carbon and low-alloy steels. All of these steels contain nominally 12% chromium and have mostly been developed from AISI type 409 ferritic stainless steel (Thomas, 1981; Lula, 1984). Design criteria for one of these steels, a steel that has been designated 3CR12, have been developed in the past (van der Merwe and Yu, 1985; van der Merwe, 1987).

7.8 CONDENSED REFERENCES / BIBLIOGRAPHY

AISI 1968, *Specification for the Design of Cold-Formed Steel Structural Members*

AISI 1974a, *Specification for the Design of Cold-Formed Stainless Steel Structural Members*

AISI 1974b, *Stainless Steel Cold-Formed Structural Design Manual*

AISI 1986, *Specification for the Design of Cold-Formed Steel Structural Members*

ASCE 1991, *Specification for the Design of Cold-Formed Stainless Steel Structural Members*

Berg 1988, *The Torsional Flexural Buckling Strength of Cold-Formed Stainless Steel Columns*

Errera 1970, *Strength of Bolted and Welded Connections in Stainless Steel*

Johnson 1966a, *The Structural Performance of Austenitic Stainless Steel Members*

Johnson 1966b, *Behavior of Stainless Steel Columns and Beams*

Johnson 1969, *Stainless Steel in Structural Applications*

Lin 1989, *Load and Resistance Factor Design of Cold-Formed Stainless Steel*

Lula 1984, *The Status and Development and Industrial Application of the 12% Chromium Steels in the United States*

Melvill 1980, *The Development of a Chromium-Containing Corrosion Resisting Steel*

Merwe 1985, *Development of Design Criteria for a Ferritic Stainless Steel*

Merwe 1987, *Development of Design Criteria for Ferritic Stainless Steel Structural Members and Connections*

Thomas 1981, *Use of a 12% Chromium Ferritic Stainless Steel in Coal Conversion Systems*

Wang 1969, *Cold-Rolled Austenitic Stainless Steel: Material Properties and Structural Performance*

Yu 1991, *Cold-Formed Steel Design*

Current Questions, Problems, and Research Needs

During the past three decades, research and development of cold-formed steel structures have been successfully conducted throughout the world. As a result, many design specifications and recommendations have been prepared in various countries for use by architects and engineers. Because new building products with unusual sectional configurations can be economically cold-formed from various grades of steels, additional research is needed to provide background information for the development of new and improved design rules.

From the engineering point of view, a large number of research projects are needed for the design of cold-formed steel members, connections, and structural systems. In view of the fact that the most commonly used cold-formed steel components in tall buildings include steel decks, curtain walls, partitions, roof systems, construction forms, purlins and joists, and duct systems, the following research subjects are related primarily to building construction:

1. General
   a. Simplification of complicated design rules
   b. Development of design criteria concerning fatigue design

2. Element behavior
   a. Effective design width of compression elements with stress gradient
   b. Behavior of long edge stiffeners
   c. Effective design width of multiple-stiffened elements
   d. Behavior and strength of perforated compression elements
   e. Effect of corner radius on effective design width

3. Member design
   a. Design rules for panels with unbraced compression flanges
   b. Design rules for decks and panels under point loading
   c. Flexural members using A446 and A611 grade E steels
   d. Additional design rules for inelastic reserve strength of beams
   e. Design rules for perforated webs
   f. Improved design rules for wall studs
   g. Improved design rules for built-up members

h. Shear buckling of webs, web crippling, and interaction with bending
i. Effective depth of webs with various flange restraints
j. Design rules for duct systems

These subjects represent only some of the research needs for the design of cold-formed steel structures. Other topics may also be found of interest for the effective and economical use of cold-formed steel in tall buildings.

Nomenclature

GLOSSARY

Blind rivet. Mechanical fastener capable of joining workpieces together where access to the assembly is limited to one side only. Installation is done by actuation of a self-contained mechanical feature which forms an upset on the blind end of the rivet and expands the rivet shank. The rivet has been installed in a predrilled hole.

Bridging. Bracing system of compressed top chord of joists.

Chord. Top or bottom element of a joist.

Cladding. Profiled sheet for walls.

Cold-forming. Operation to shape from steel sheet, strip, or plate by means of roll-forming machines or press brake.

Composite slab. Floor in which the structural bearing capacity is formed by the cooperation of concrete and floor decking (as reinforcement).

Deck. Number of deckings assembled in series.

Decking. Profiled sheet for floors or roofs.

Effective length. Of a fillet weld, the overall length including end returns.

Embossments. Deformations (indentures) cold-rolled into decking to provide mechanical interaction between concrete and steel.

Monel. Alloy of mainly nickel and copper as material to produce blind rivets.

Panel. Profiled sheet for floors, roofs, or walls.

Primary force. Force directly caused by load.

Props. Temporary supports under decking to provide sufficient strength and stiffness during concreting and hardening of the concrete.

Prying force. Extra tension force in a fastener generated by an accidental fixing moment.

Quality profile. Set of specific quality requirements for a building element applied in a selected type of building.

Sandwich panel. Load-bearing element consisting of two metal skins with a thick core having low strength and density.

Secondary force. Force indirectly caused by the load and which may be neglected in the presence of sufficient deformation capacity in the fastening.

Self-drilling screw. Screw that drills its own hole and forms its mating thread in one operation.

Self-tapping screw. Screw that taps a counterthread in a prepared hole.

Sheet. Flat or profiled metal layer.

Sheeting. Profiled sheet for floors, roofs, or walls.

Shot pin. Fastener driven through the material to be fastened into the base metal structure. The driving energy can be provided by propellant or compressed air.

Shuttering. Formwork for concreting.

Steel core thickness. Thickness of a sheet exclusive of the coating (zinc or paint) thickness.

Steel joist. Open-web type beam to support floors or roofs.

Structural member. Element intended for bearing loads or forces.

Web crippling. Failure mode of the web of a beam caused by the combination of a bending moment and a concentrated load (such as a support reaction).

Wrinkling. Failure mode of the compressed skin of a sandwich panel which is loaded in bending, appearing as buckles having a wave length of the same order as the core thickness.

SYMBOLS

The numerals in parentheses refer to the chapter in which the given symbol is used.

A_c = compression chord cross-sectional area (2)
A_{et} = tensile stress area of bolt (5)
A_g = area of gross cross-section (2)
A_n = net cross-sectional area of plate material (2, 5)
A_s = spandrel beam area (4)
A_t = tension chord cross-sectional area (2)
B = building shear width (4)
b = width of specimen (5)
b_e = effective width of a cross-section in compression board material
C_c = column slenderness ratio separating elastic and inelastic buckling (2)
$c_{1.1}$ = flexibility due to distortion of deck profile (4)
$c_{1.2}$ = flexibility due to shear strain in sheet (4)
$c_{2.1}$ = flexibility due to fastener slip along seams (4)
$c_{2.2}$ = flexibility due to fastener slip along diaphragm edges (4)
$c_{2.3}$ = flexibility due to fastener slip perpendicular to deck span (4)
c_3 = flexibility due to flexural action of equivalent plate girder (4)
D = decking depth (4)
d = distance between compression and tension chord centroids (2)
= visible weld diameter (4)
= screw diameter (4)
= nominal diameter of hole (5)
d_0 = washer hole diameter (4)
d_{eff} = effective diameter of circular plug weld (5)
d_n = nominal diameter of fastener (5)
d_s = diameter of spot weld (5)

d_w	=	diameter of washer or head of fastener	(5)
	=	visible diameter of circular plug weld	(5)
E	=	Young's modulus	(4)
E_0	=	initial modulus of elasticity	(7)
E_c	=	mean Young's modulus of core	(3)
E_i	=	Young's modulus of each core layer	(3)
E_r	=	reduced modulus of elasticity	(7)
E_s	=	secant modulus of elasticity at unfactored stresses	(7)
E_{sc}	=	secant modulus of elasticity in compression	(7)
E_{st}	=	secant modulus of elasticity in tension	(7)
E_t	=	tangent modulus of elasticity at unfactored stresses	(7)
e	=	end distance of spot weld	(5)
e_1	=	edge distance in load direction	(5)
e_2	=	center-to-center distance in load direction	(5)
F	=	stress function	(2)
	=	force in a fastening caused by design load	(5)
F_s	=	shear force developed in fastner	(4)
F^*	=	characteristic shear capacity of single spot weld with subscript depending on failure mode (Table 5.5)	(5)
F_a	=	Allowable compression stress	(2)
F_b^*	=	characteristic shear strength of a connection per fastener, failure mode hole bearing (including tilting and shear of section)	(5)
F_c	=	lateral-torsional buckling stress	(2)
	=	allowable compression stress for joists	(2)
F_d	=	design strength of a fastening	(5)
F_e	=	shear strength of end fasteners	(4)
F_n^*	=	characteristic strength of a connection, failure mode yield of net section	(5)
F_o^*	=	characteristic pullout strength of a connection per fastener	(5)
F_p^*	=	characteristic pull-through–pull-over strength of a connection per fastener	(5)
F_r	=	resulting force in corner fastener	(4)
F_s	=	shear in seam fastener	(4)
F_s^*	=	calculated characteristic shear capacity of spot weld	(5)
F_{st}	=	shear in a shear transfer fastener	(4)
F_t	=	allowable tension stress for joists	(2)
F_u	=	tensile strength of steel	(2, 4)
F_{ue}	=	ultimate strength of end fastener	(4)
F_{uw}	=	ultimate strength of interior fastener	(4)
F_v^*	=	characteristic capacity per bolt per shear plan (for thin-walled steel thread is always up to head)	(5)

F_w = shear in interior fastener (4)
F_{wb} = web bending stress (2)
F_{xx} = electrode strength (4)
F_y = yield strength of steel (2)
f_c = concrete compressive strength (4)
f_n = design value for net section stress (5)
f_u = specified ultimate tensile strength of steel sheet with thickness t (5)
= design strength of steel (5)
f_{ub} = specified ultimate tensile strength of bolt material (5)
f_w = design strength of weld material (5)
f_y = yield strength of steel (5)
g = deformation in plane of decking (3)
G = profile length (3)
G' = diaphragm shear stiffness (4)
G_0 = initial shear modulus (7)
G_s = secant shear modulus at unfactored stresses (7)
H = web slenderness ratio (2)
h = profile height (3)
h_1 = deflection caused by difference in lengths between straight and curved neutral axes (5)
h_2 = deflection caused by distance between neutral axis and flange through which it is fastened (5)
I = moment of inertia of a joist (2)
= beam moment of inertia (4)
= moment of inertia of a sump pan (3)
K = fastner stiffness (4)
= column effective length ratio (2)
K_e = stiffness of end fastener (4)
K_s = stiffness of seam fastener (4)
K_{st} = stiffness of shear transfer fastener (4)
K_w = stiffness of interior fastener (4)
L = span of floor structure (6)
= building length (4)
L/r = slenderness ratio for chord and web elements in joists (2)
L_s = decking length (4)
L_v = purlin spacing (4)
= length of deck span (4)
L_w = length of elongated plug weld (5)
l = span of the deck (3)
l_w = effective length of fillet weld (5)

M = resisting moment of joist (2)
= moment (3)
M_e = end of decking couple (4)
M_p = interior purlin couple (4)
M_r = factored resistance for member in bending (2)
M_{ult} = ultimate failure moment (2)
M_{yield} = yield moment (2)
n_p = number of interior purlins (4)
= number of structural supports (4)
n_s = number of seam fasteners along one seam of deck sheet (4)
n_{st} = number of shear transfer connections along entire edge of deck sheet (4)
P = seam strength (4)
P_u = ultimate shear load (4)
Q = section effective area ratio (2)
Q_f = fastner shear strength (4)
Q_s = stitch connector shear strength (4)
q = uniformly distributed load (4, 6)
R = specified nominal value of resistance (2)
= interior wall reactions (4)
= characteristic shear capacity of a spot weld according to prescribed test procedure (5)
$\bar{R}$ = mean value of actual resistance (2)
r = force transmitted by bolt or bolts at section considered, divided by tension force in member at that section (5)
S = specified nominal value of load effects; section modulus (2)
$\bar{S}$ = mean value of actual load effects (2)
S_u = average ultimate shear strength per unit length (4)
SF = safety factor
SP = section modulus of sump pan (3)
s = spacing of seam fasteners (4)
T_r = factored resistance for member in tension (2)
t = thickness of thinnest sheet (4, 5)
= thickness of plate material (5)
= steel core thickness (3)
t_1 = thickness of thickest sheet (base material) (5)
t_c = thickness of core (3)
t_f = thickness of face (3)
t_i = thickness of each layer of core (3)
t_{si} = equivalent value of steel thickness (6)

u = difference between loads and resistance in ln values (2)

$\bar{u}$ = mean value of differences between loads and resistance in ln values (2)

u_1 = distance between edge and center of fastener perpendicular to load direction (5)

u_2 = center-to-center spacing of fasteners perpendicular to load direction (5)

V = applied shear per unit length (4)

= shear force in diaphragm (4)

V_L = reaction of left support (Fig. 4.1) (4)

V_R = reaction of right support (Fig. 4.1) (4)

W = flat-width ratio of flange (2)

W_1 = wet concrete plus deck weight, psf (3)

W_2 = construction load of 20 psf (3)

w = decking width (4)

= concrete unit weight (4)

x_i = distance from centerline of deck to transverse fasteners (end or interior fasteners) (4)

z = number of welds across deck width (4)

α = aggregate load factor (2)

= relative stiffness (4)

= number of shear transfer elements between joists (4)

= factor defined at relevant place in Table 5.3 (5)

β = safety index, reliability index (2)

= tension-to-compression stress ratio (2)

= number of continuous spans (4)

γ_m = material factor (5)

Δ = deformation of fastener under load (4)

δ = midspan deflection, including ponding effect caused by self-weight of decking and concrete (3)

ϕ = diameter of mechanical fastener (5)

ϕ_u = resistance factor for tensile strength limit state of material (2)

ρ_f = density of thin face (3)

Φ = resistance factor (2)

ABBREVIATIONS

AISC American Institute of Steel Construction

AISI American Iron and Steel Institute

ASCE American Society of Civil Engineers

ASD Allowable stress design

ASTM	American Society for Testing and Materials
AWS	American Welding Society
BS	British Standard
CEN	Convention Européenne de Normalisation (European Committee for Standardization)
CSA	Canadian Standards Association
ECCS	European Convention for Constructional Steelwork
FMC	Factory Mutual Corporation
IABSE	International Association for Bridge and Structural Engineering
IISI	International Iron and Steel Institute
LRFD	Load and resistance factor design
LSD	Limit state design
NBCC	National Building Code of Canada
prEN	Preliminary European Norm
RMI	Rack Manufacturers Institute
SCI	Steel Construction Institute
SDI	Steel Deck Institute
SIS	Samenwerkingsverband Industrie Staalplaat-betonvloeren (Association for the Industry of Composite Slabs)
SJI	Steel Joist Institute
SSRC	Structural Stability Research Council
TNO	Toegepast Natuurwetenschappelijk Onderzoek (Applied Scientific Research)

UNITS

In the table below are given conversion factors for commonly used units. The numerical values have been rounded off to the values shown. The British (Imperial) System of units is the same as the American System except where noted. Le Système International d'Unités (abbreviated "SI") is the name formally given in 1960 to the system of units partly derived from, and replacing, the old metric system.

SI	American	Old metric
	Length	
1 mm	0.03937 in.	1 mm
1 m	3.28083 ft 1.093613 yd	1 m
1 km	0.62137 mile	1 km

SI	American	Old metric
	Area	
1 mm^2	0.00155 $in.^2$	1 mm^2
1 m^2	10.76392 ft^2	1 m^2
	1.19599 yd^2	
1 km^2	247.1043 acres	1 km^2
1 hectare	2.471 acres(1)	1 hectare
	Volume	
1 cm^3	0.061023 $in.^3$	1 cc
		1 ml
1 m^3	35.3147 ft^3	1 m^3
	1.30795 yd^3	
	264.172 gal(2) liquid	
	Velocity	
1 m/sec	3.28084 ft/sec	1 m/sec
1 km/hr	0.62137 mile/hr	1 km/hr
	Acceleration	
1 m/sec^2	3.28084 ft/sec^2	1 m/sec^2
	Mass	
1 g	0.035274 oz	1 g
1 kg	2.2046216 lb(3)	1 kg
	Density	
1 kg/m^3	0.062428 lb/ft^3	1 kg/m^3
	Force, Weight	
1 N	0.224809 lbf	0.101972 kgf
1 kN	0.1124045 ton(4)	
1 MN	224.809 kips	
1 kN/m	0.06853 kips/ft	
1 kN/m^2	20.9 lbf/ft^2	
	Torque, Bending Moment	
1 N-m	0.73756 lbf-ft	0.101972 kgf-m
1 kN-m	0.73756 kip-ft	101.972 kgf-m
	Pressure, Stress	
1 N/m^2 = 1 Pa	0.000145038 psi	0.101972 kgf/m^2
1 kN/m^2 = 1 kPa	20.8855 psf	
1 MN/m^2 = 1 MPa	0.145038 ksi	
	Viscosity (Dynamic)	
1 $N\text{-}sec/m^2$	0.0208854 $lbf\text{-}sec/ft^2$	0.101972 $kgf\text{-}sec/m^2$

SI	American	Old metric
	Viscosity (Kinematic)	
1 m^2/sec	10.7639 ft^2/sec	1 m^2/sec
	Energy, Work	
1 J = 1 N-m 1 MJ	0.737562 lbf-ft 0.37251 hp-hr	0.00027778 w-hr 0.27778 kw-hr
	Power	
1 W = 1 J/sec 1 kW	0.737562 lbf ft/sec 1.34102 hp	1 w 1 kw
	Temperature	
K = 273.15 + °C K = 273.15 + 5/9(°F − 32) K = 273.15 + 5/9(°R − 491.69)	°F = (°C × 1.8) + 32	°C = (°F − 32)/1.8

(1) Hectare as an alternative for km^2 is restricted to land and water areas.
(2) 1 m^3 = 219.9693 Imperial gallons.
(3) 1 kg = 0.068522 slugs.
(4) 1 American ton = 2000 lb. 1 kN = 0.1003612 Imperial ton. 1 Imperial ton = 2240 lb.

Abbreviations for Units

Btu	British thermal unit	kW	kilowatt
°C	degree Celsius (centigrade)	lb	pound
cc	cubic centimeters	lbf	pound force
cm	centimeter	lb_m	pound mass
°F	degree Fahrenheit	MJ	megajoule
ft	foot	MPa	megapascal
g	gram	m	meter
gal	gallon	ml	milliliter
hp	horsepower	mm	millimeter
hr	hour	MN	meganewton
Imp	British Imperial	N	newton
in.	inch	oz	ounce
J	joule	Pa	pascal
K	Kelvin	psf	pounds per square foot
kg	kilogram	psi	pounds per square inch
kgf	kilogram-force	°R	degree Rankine
kip	1000 pound force	sec	second
km	kilometer	slug	14.594 kg
kN	kilonewton	U_o	heat transfer coefficient
kPa	kilopascal	W	watt
ksi	kips per square inch	yd	yard

References/Bibliography

The citations that follow include both references and bibliography. The list includes all articles cited in the text, and it also includes a bibliography for further reading. Additional bibliographies are available through the Council.

Academia Sinica, 1977
BENDING, STABILITY AND VIBRATION OF SANDWICH PANEL (in Chinese), Solid Mechanics Group, Beijing, China.

Ahmed, K. M., 1971
FREE VIBRATION OF CURVED SANDWICH PANELS BY THE METHOD OF FINITE ELEMENTS, *Journal of Sound and Vibration,* vol. 18, pp. 61–74.

AISC, 1980
MANUAL OF STEEL CONSTRUCTION, 8th edition, American Institute of Steel Construction, Chicago, Ill.

AISC, 1989a
SPECIFICATION FOR THE DESIGN, FABRICATION AND ERECTION OF STRUCTURAL STEEL FOR BUILDINGS, American Institute of Steel Construction, Chicago, Ill.

AISC, 1989b
MANUAL OF STEEL CONSTRUCTION, 9th edition, American Institute of Steel Construction, Chicago, Ill.

AISI, 1967
DESIGN OF SHEAR DIAPHRAGMS, American Iron and Steel Institute, Washington, D.C.

AISI, 1968
SPECIFICATION FOR THE DESIGN OF COLD-FORMED STEEL STRUCTURAL MEMBERS, American Iron and Steel Institute, Washington, D.C.

AISI, 1974a
SPECIFICATION FOR THE DESIGN OF COLD-FORMED STAINLESS STEEL STRUCTURAL MEMBERS, American Iron and Steel Institute, Washington, D.C.

AISI, 1974b
STAINLESS STEEL COLD-FORMED STRUCTURAL DESIGN MANUAL, American Iron and Steel Institute, Washington, D.C.

AISI, 1980
SPECIFICATION FOR THE DESIGN OF COLD-FORMED STEEL STRUCTURAL MEMBERS, American Iron and Steel Institute, Washington, D.C.

AISI, 1986
SPECIFICATION FOR THE DESIGN OF COLD-FORMED STEEL STRUCTURAL MEMBERS, August 19, 1986, and Addendum, December 11, 1989, American Iron and Steel Institute, Washington, D.C.

AISI, 1991a
LOAD AND RESISTANCE FACTOR DESIGN SPECIFICATION FOR COLD-FORMED STEEL STRUCTURAL MEMBERS, American Iron and Steel Institute, Washington, D.C.

AISI, 1991b
COMMENTARY ON THE LOAD AND RESISTANCE FACTOR DESIGN SPECIFICATION FOR COLD-FORMED STEEL STRUCTURAL MEMBERS, American Iron and Steel Institute, Washington, D.C.
Albrecht, R. E., 1989
USE OF COLD-FORMED STEEL AS COMPOSITE FLOOR DECK IN TALL BUILDINGS, private correspondence.
Allen, D. E., 1975
LIMIT STATES DESIGN—A PROBABILISTIC STUDY, *Canadian Journal of Civil Engineering,* vol. 2, March.
Allen, D. E., 1981
LIMIT STATES DESIGN: WHAT DO WE REALLY WANT, *Canadian Journal of Civil Engineering,* vol. 8, March.
Allen, H. G., 1969
ANALYSIS AND DESIGN OF STRUCTURAL SANDWICH PANELS, New York: Pergamon Press.
Architectural Record, 1976
LIGHT-GAGE STEEL IN THE FRAMING FOR LIGHTWEIGHT WALL PANELS, New York: McGraw-Hill, pp. 125–126.
ASCE, 1969
LITERATURE SURVEY OF COLD-FORMED STRUCTURES, ASCE Subcommittee on Literature Survey of Cold-Formed Structures, K. P. Chong, chairman, ASCE Convention, Atlanta, Ga., preprint 3762.
ASCE, 1984
SPECIFICATION FOR THE DESIGN AND CONSTRUCTION OF COMPOSITE SLABS AND COMMENTARY ON SPECIFICATIONS FOR THE DESIGN AND CONSTRUCTION OF COMPOSITE SLABS, American Society of Civil Engineers, New York.
ASCE, 1991
SPECIFICATION FOR THE DESIGN OF COLD-FORMED STAINLESS STEEL STRUCTURAL MEMBERS, ANSI/ASCE-8-90, American Society of Civil Engineers, New York.
Atrek, E., and Nilson, A. H., 1980
NONLINEAR ANALYSIS OF COLD-FORMED STEEL SHEAR DIAPHRAGMS, *Journal of the Structural Division, ASCE,* vol. 106, pp. 693–710.
AWS, 1981
STRUCTURAL WELDING CODE, SHEET STEEL, D 1.3, American Welding Society.
Badoux, J. C., and Crisinel, M., 1973
RECOMMENDATIONS FOR THE APPLICATION OF COLD-FORMED STEEL DECKING FOR COMPOSITE SLABS IN BUILDINGS, Swiss Institute of Steel Construction.
Baehre, R., 1978
DEVELOPMENT CHARACTERISTICS OF THIN-WALLED BUILDING TECHNICS—SECTION STIFFENING, COMPONENTS, COMPOSITE ACTION, Document D8: 1978, Swedish Council for Building Research, Stockholm.
Baehre, R., 1982
SPACE COVERING BUILDING ELEMENTS (in German), Cover 1, Chapter 17, Stahlbau-Verlagsgesellschaft, Köln, Germany.
Baehre, R., Bryan, E. R., He, B. K., Moreau, G., Pekoz, T., and Sakae, K., 1982
COLD-FORMED STEEL APPLICATIONS ABROAD, Proceedings of 6th International Specialty Conference on Cold-Formed Steel Structures, University of Missouri-Rolla, Rolla, Mo., pp. 613–652.
Baehre, R., and Urschel, H., 1984
LIGHT-WEIGHT STEEL BASED FLOOR SYSTEMS FOR MULTI-STORY BUILDINGS, Proceedings of 7th International Specialty Conference on Cold-Formed Steel Structures, University of Missouri-Rolla, Rolla, Mo., pp. 375–394.

Balasz, P., 1980
STRESSED SKIN ACTION IN COMPOSITE PANELS COMPRISING STEEL SHEETING AND BOARDS, Document D40: 1980, Swedish Council for Building Research, Stockholm.
Balasz, P., and Thomasson, P. O., 1985
TESTS ON BOX UNITS OF STEEL SHEETING IN PARTIAL STRUCTURE INTERACTION WITH PLASTER BOARD, Document D23: 1985, Swedish Council for Building Research, Stockholm.
Beedle, L. S., Editor-in-Chief, 1991
STABILITY OF METAL STRUCTURES, A WORLD VIEW, 2d edition, Structural Stability Research Council, Bethlehem, Pa.
Berg, G. J. van den, 1988
THE TORSIONAL FLEXURAL BUCKLING STRENGTH OF COLD-FORMED STAINLESS STEEL COLUMNS, D.Eng. Thesis, Rand Afrikaans University, Johannesburg, Republic of South Africa.
Berner, K., and Stemmann, D., 1990
SANDWICH-PANELS WITH STEEL FACINGS AND DIFFERENT CORE MATERIALS, Proceedings of IABSE Conference, Brussels, Belgium, September, pp. 551–556.
Bode, H., Kunzel, R., and Schanzenbach, J., 1988
PROFILED STEEL SHEETING AND COMPOSITE ACTION, Proceedings of 9th International Specialty Conference on Cold-Formed Steel Structures, University of Missouri-Rolla, Rolla, Mo., pp. 343–360.
Brekelmans, J. W. P. M., O'Leary, D. C., and Moum, Ch., 1990
COMPARATIVE STUDY OF COMPOSITE SLAB TESTS, Proceedings of IABSE Conference, Brussels, Belgium, September.
Bryan, E. R., 1972
THE STRESSED SKIN DESIGN OF STEEL BUILDINGS, Crosby Lockwood Staples, London.
Bryan, E. R., and Davies, J. M., 1981
STEEL DIAPHRAGM ROOF DECKS, A DESIGN GUIDE WITH TABLES FOR ENGINEERS AND ARCHITECTS, Granada Publishing, London.
Bryan, E. R., and Davies, J. M., 1982
MANUAL OF STRESSED SKIN DIAPHRAGM DESIGN, Granada Publishing, London.
Bryan, E. R., Sedlacek, G., Tomà, A. W., and Weynand, K., 1990
EVALUATION OF TEST RESULTS ON CONNECTIONS IN THIN WALLED SHEETING AND MEMBERS IN ORDER TO OBTAIN STRENGTH FUNCTIONS AND SUITABLE MODEL FACTORS, Background Document A.01 to Annex A of Eurocode 3, Institute of Steel Construction, Aachen, Germany.
BS 5950: Part 5, 1987
STRUCTURAL USE OF STEELWORK IN BUILDING, PART 5: CODE OF PRACTICE FOR DESIGN OF COLD FORMED SECTIONS, British Standard, London.
Bucheli, P., and Crisinel, M., 1982
COMPOSITE BEAMS IN BUILDINGS (in French), Swiss Institute of Steel Construction.
CAN3-S136-1974, 1974
COLD FORMED STEEL STRUCTURAL MEMBERS, Canadian Standards Association, Rexdale, Ont.
CAN3-S136-M84, 1984
COLD FORMED STEEL STRUCTURAL MEMBERS, Canadian Standards Association, Rexdale, Ont.
CAN3-S136-M89, 1989
COLD FORMED STEEL STRUCTURAL MEMBERS, Canadian Standards Association, Rexdale, Ont.

Canadian Sheet Steel Building Institute, 1968
COMPOSITE BEAM MANUAL FOR THE DESIGN OF STEEL BEAMS WITH CONCRETE SLAB AND CELLULAR STEEL FLOOR, Canadian Sheet Steel Building Institute, Willowdale, Ont.

Canadian Sheet Steel Building Institute, 1986
STANDARDS FOR STEEL ROOF DECK, CSSBI 10M-86, Canadian Sheet Steel Building Institute, Willowdale, Ont., March.

Cheung, Y. K., 1968a
FINITE STRIP METHOD IN THE ANALYSIS OF ELASTIC PLATES WITH TWO OPPOSITE SIMPLY SUPPORTED ENDS, *Proceedings of the Institution of Civil Engineers,* vol. 40, pp. 1–7.

Cheung, Y. K., 1968b
FINITE STRIP METHOD ANALYSIS OF ELASTIC SLABS, *Journal of Engineering Mechanics, ASCE,* vol. 94, pp. 1365–1378.

Cheung, Y. K., 1976
FINITE STRIP METHOD IN STRUCTURAL ANALYSIS, Pergamon Press, New York.

Cheung, Y. K., 1981
FINITE STRIP METHOD IN STRUCTURAL AND CONTINUUM MECHANICS, *International Journal of Structures,* vol. 1, no. 1, pp. 19–37.

Cheung, Y. K., Chong, K. P., and Tham, L. G., 1982
BUCKLING OF SANDWICH PLATE BY FINITE LAYER METHOD, *Computers and Structures,* vol. 15, no. 2, pp. 131–134.

Cheung, Y. K., Yeo, M. F., and Cumming, D. A., 1976
THREE-DIMENSIONAL ANALYSIS OF FLEXIBLE PAVEMENTS WITH SPECIAL REFERENCE TO EDGE LOADS, Proceedings of 1st Conference of the Road Engineering Association of Asia and Australia, Bangkok, Thailand.

Chong, K. P., 1986
SANDWICH PANELS WITH COLD-FORMED THIN FACINGS, Proceedings of IABSE Colloquium, Stockholm, Sweden, May, pp. 339–348.

Chong, K. P., Cheung, Y. K., and Tham, L. G., 1982a
FREE VIBRATION OF FOAMED SANDWICH PANEL, *Journal of Sound and Vibration,* vol. 81, pp. 575–582.

Chong, K. P., Engen, K. O., and Hartsock, J. A., 1976
THERMAL STRESS IN DETERMINATE AND INDETERMINATE SANDWICH PANELS WITH FORMED FACINGS, ASCE-EMD, 1st Specialty Conference, Waterloo, Ont., Canada.

Chong, K. P., Engen, K. O., and Hartsock, J. A., 1977
THERMAL STRESS AND DEFLECTION OF SANDWICH PANELS, *Journal of the Structural Division, ASCE,* pp. 35–49.

Chong, K. P., and Hartsock, J. A., 1972
FLEXURAL WRINKLING MODE OF ELASTIC BUCKLING IN SANDWICH PANELS, Proceedings of ASCE Specialty Conference on Composite Materials, Pittsburgh, Pa.

Chong, K. P., and Hartsock, J. A., 1974
FLEXURAL WRINKLING IN FOAM-FILLED SANDWICH PANELS, *Journal of the Engineering Mechanics Division, ASCE,* pp. 95–110.

Chong, K. P., Lee, B., and Lavdas, P. A., 1984
ANALYSIS OF THIN-WALLED STRUCTURES BY FINITE STRIP AND FINITE LAYER METHODS, *Thin-Walled Structures,* vol. 2, pp. 75–95.

Chong, K. P., Tham, L. G., and Cheung, Y. K., 1982b
THERMAL BEHAVIOR OF FOAMED SANDWICH PLATE BY FINITE-PRISM-STRIP METHOD, *Computers and Structures,* vol. 15, no. 3, pp. 321–324.

Chong, K. P., Wang, K. A., and Griffith, G. R., 1979
ANALYSIS OF CONTINUOUS SANDWICH PANELS IN BUILDING SYSTEM, *Building and Environment,* vol. 14, pp. 125–130.

Cohn, B. M., TD4, 1972
DESIGN OF FIRE-RESISTIVE ASSEMBLIES WITH STEEL JOISTS, Technical Digest 4, Steel Joist Institute, Myrtle Beach, S.C.

Council on Tall Buildings, 1983
DEVELOPMENTS IN TALL BUILDINGS 1983, First updating of *Monograph on Planning and Design of Tall Buildings,* Hutchinson Ross, Stroudsburg, Pa.
Council on Tall Buildings, 1986
ADVANCES IN TALL BUILDINGS, Second updating of *Monograph on Planning and Design of Tall Buildings,* Hutchinson Ross, Stroudsburg, Pa.
Council on Tall Buildings, 1986
HIGH-RISE BUILDINGS: RECENT PROGRESS, Third updating of *Monograph on Planning and Design of Tall Buildings,* Council on Tall Buildings and Urban Habitat, Bethlehem, Pa.
Council on Tall Buildings, Group CB, 1978
STRUCTURAL DESIGN OF TALL CONCRETE AND MASONRY BUILDINGS, Volume CB of *Monograph on Planning and Design of Tall Buildings,* ASCE, New York.
Council on Tall Buildings, Group CL, 1980
TALL BUILDING CRITERIA AND LOADING, Volume CL of *Monograph on Planning and Design of Tall Buildings,* ASCE, New York.
Council on Tall Buildings, Group PC, 1981
PLANNING AND ENVIRONMENTAL CRITERIA STRUCTURAL FOR TALL BUILDINGS, Volume SB of *Monograph on Planning and Design of Tall Buildings,* ASCE, New York.
Council on Tall Buildings, Group SB, 1979
STRUCTURAL DESIGN OF TALL STEEL BUILDINGS, Volume SB of *Monograph on Planning and Design of Tall Buildings,* ASCE, New York.
Council on Tall Buildings, Group SC, 1980
TALL BUILDING SYSTEMS AND CONCEPTS, Volume SC of *Monograph on Planning and Design of Tall Buildings,* ASCE, New York.
Cran, J. A., 1971
DESIGN AND TESTING COMPOSITE OPEN WEB STEEL JOISTS, Proceedings of 1st International Specialty Conference on Cold-Formed Steel Structures, University of Missouri-Rolla, Rolla, Mo., pp. 186–197.
CSA-S408-1981, 1981
GUIDELINES FOR THE DEVELOPMENT OF LIMIT STATES DESIGN, Special Publication, Canadian Standards Association, Rexdale, Ont.
Daniels, B. J., 1990
BEARING CAPACITY OF COMPOSITE SLABS: MATHEMATICAL MODELING AND EXPERIMENTAL STUDY (in French), Doctoral thesis no. 895, Ecole Polytechnique Fédérale de Lausanne, Switzerland.
DASt-Richtlinie 016, 1988
CALCULATION AND DESIGN OF STRUCTURES FROM COLD-FORMED BUILDING ELEMENTS (in German), Stahlbau-Verlagsgesellschaft, Köln, Germany.
Davies, J. M., 1977
SIMPLIFIED DIAPHRAGM ANALYSIS, *Journal of the Structural Division, ASCE,* vol. 103, pp. 2093–2110.
Davies, J. M., and Hakmi, M. R., 1990
LOCAL BUCKLING OF PROFILED SANDWICH PLATES, Proceedings of IABSE Conference, Brussels, Belgium, September, pp. 533–538.
DeWolf, J. T., 1973
LOCAL AND OVERALL BUCKLING OF COLD-FORMED COMPRESSION MEMBERS, Report 354, Cornell University, Ithaca, N.Y.
Easeley, J. T., 1977
STRENGTH AND STIFFNESS OF CORRUGATED METAL SHEAR DIAPHRAGMS, *Journal of the Structural Division, ASCE,* vol. 103, pp. 169–180.
Easterling, W. S., and Porter, M. L., 1988
COMPOSITE DIAPHRAGM BEHAVIOR AND STRENGTH, Proceedings of 9th International Specialty Conference on Cold-Formed Steel Structures, University of Missouri-Rolla, Rolla, Mo., pp. 387–404.

ECCS, 1975
EUROPEAN RECOMMENDATIONS FOR THE CALCULATION AND DESIGN OF COMPOSITE FLOORS WITH PROFILED STEEL SHEET, ECCS Committee 11, ECCS, Brussels, Belgium.
ECCS, 1977
EUROPEAN RECOMMENDATIONS FOR THE STRESSED SKIN DESIGN OF STEEL STRUCTURES, ECCS Committee 17, ECCS, Brussels, Belgium.
ECCS, 1981
COMPOSITE STRUCTURES, MODEL CODE, Joint Committee on Composite Structures, Harlow, Essex, U.K.: The Construction Press.
ECCS, 1983a
EUROPEAN RECOMMENDATIONS FOR THE DESIGN OF PROFILED SHEETING, ECCS Committee TC7, ECCS, Brussels, Belgium, April.
ECCS, 1983a
THE DESIGN AND TESTING OF CONNECTIONS IN STEEL SHEETING AND SECTIONS, ECCS, Brussels, Belgium, May.
ECCS, 1983b
EUROPEAN RECOMMENDATIONS FOR GOOD PRACTICE IN STEEL CLADDING AND DECKING, ECCS Committee TC7, ECCS, Brussels, Belgium, May.
ECCS, 1983b
MECHANICAL FASTENERS FOR USE IN STEEL SHEETING AND SECTIONS, ECCS, Brussels, Belgium, June.
ECCS, 1984
LIGHTWEIGHT STEEL BASED FLOOR SYSTEMS FOR MULTI-STOREY BUILDINGS, ECCS Advisory Committee 1, Publication 34, ECCS, Brussels, Belgium.
ECCS, 1987
EUROPEAN RECOMMENDATIONS FOR THE DESIGN OF LIGHT GAUGE STEEL MEMBERS, ECCS Committee TC7, Publication 49, ECCS, Brussels, Belgium.
ECCS, 1990a
EUROPEAN RECOMMENDATIONS FOR SOUND INSULATION OF STEEL CONSTRUCTION IN MULTI-STORY BUILDINGS, ECCS Advisory Committee AC1, Publication 59, ECCS, Brussels, Belgium.
ECCS, 1990b
EUROPEAN RECOMMENDATIONS FOR SANDWICH PANELS, PART II: GOOD PRACTICE, ECCS Committee TC7, Publication 62, ECCS, Brussels, Belgium.
ECCS, 1991
EUROPEAN RECOMMENDATIONS FOR SANDWICH PANELS, PART I: DESIGN, Draft, ECCS Committee TC7, ECCS, Brussels, Belgium, May.
Ekberg, C. E., and Schuster, R. M., 1968
FLOOR SYSTEMS WITH COMPOSITE FORM-REINFORCED CONCRETE SLABS, Final Report, 8th Congress, International Association for Bridge and Structural Engineering.
Errera, S. J., Tang, B. M., and Popowich, D. W., 1970
STRENGTH OF BOLTED AND WELDED CONNECTIONS IN STAINLESS STEEL, Report 335, Department of Structural Engineering, Cornell University, Ithaca, N.Y.
Eurocode 3 Annex A, 1991
COLD FORMED STEEL SHEETING AND MEMBERS, Draft, CEN, Brussels, Belgium, January.
Eurocode 4 Part 1, 1990
DESIGN OF COMPOSITE STEEL AND CONCRETE STRUCTURES, PART 1— GENERAL RULES AND RULES FOR BUILDINGS, Draft, CEN, Brussels, Belgium, October.

Eurocode 4 Part 10, 1990
DESIGN OF COMPOSITE STEEL AND CONCRETE STRUCTURES, PART 10—STRUCTURAL FIRE DESIGN, Draft, CEN, Brussels, Belgium, April.

Euronorm 148, 1979
CONTINUOUSLY HOT-DIP ZINC COATED UNALLOYED MILD STEEL SHEET AND COIL WITH SPECIFIED MINIMUM YIELD STRENGTHS FOR STRUCTURAL PURPOSES; TOLERANCES ON DIMENSION AND SHAPE, CEN, Brussels, Belgium.

Fisher, J. W., 1970
DESIGN OF COMPOSITE BEAMS WITH FORMED METAL DECK, *Engineering Journal,* American Institute of Steel Construction, Chicago, Ill., pp. 88–96.

Fox, S. R., 1983
PREDICTING THE PRONENESS OF BUILDINGS TO GROSS ERRORS IN DESIGN AND CONSTRUCTION, M.A.Sc. Thesis, Department of Civil Engineering, University of Waterloo, Waterloo, Ont., Canada.

Fox, S. R., and Yates, D., 1986
A DESIGN METHOD FOR STEEL DECK SHEAR DIAPHRAGMS, Workshop Proceedings, 3d International Conference on Tall Buildings, Council Report M370, January 6.

Galambos, T. V., 1970a
DESIGN OF COMPRESSION CHORDS FOR OPEN WEB STEEL JOISTS, Technical Digest 1, Steel Joist Institute, Myrtle Beach, S.C., January.

Galambos, T. V., 1970b
SPACING OF BRIDGING FOR OPEN WEB STEEL JOISTS, Technical Digest 2, Steel Joist Institute, Myrtle Beach, S.C., September.

Galambos, T. V., 1978
STRUCTURAL DESIGN OF STEEL JOIST ROOFS TO RESIST UPLIFT LOADS, Technical Digest 6, Steel Joist Institute, Myrtle Beach, S.C., August.

Galambos, T. V., TD5, 1988.
VIBRATION OF STEEL JOISTS—CONCRETE SLAB FLOORS (Revised), Technical Digest 5, Steel Joist Institute, Myrtle Beach, S.C.

Gough, C. S., Elam, C. F., and Bruyne, N. A. de, 1940
THE STABILISATION OF A THIN SHEET BY A CONTINUOUS SUPPORTING MEDIUM, *Journal of the Royal Aeronautical Society,* vol. 44, pp. 12–43.

Grant, J. A., Fisher, J. W., and Slutter, R. G., 1977
COMPOSITE BEAMS WITH FORMED STEEL DECK, *Engineering Journal,* American Institute of Steel Construction, Chicago, Ill., First Quarter, pp. 24–43.

Harris, B. J., and Nordby, G. M., 1969
LOCAL FAILURE OF PLASTIC-FOAM CORE SANDWICH PANELS, *Journal of the Structural Division, ASCE,* vol. 95, pp. 585–610.

Hartsock, J. A., 1969
DESIGN OF FOAM-FILLED STRUCTURES, Technomic Publishing, Stamford, Conn.

Hartsock, J. A., and Chong, K. P., 1976
ANALYSIS OF SANDWICH PANELS WITH FORMED FACES, *Journal of the Structural Division, ASCE,* vol. 102, pp. 803–819.

Hassinen, P., 1986
COMPRESSION STRENGTH OF THE PROFILED FACE IN SANDWICH PANELS, Proceedings of IABSE Colloquium, Stockholm, Sweden, May, pp. 365–372.

Hassinen, P., and Helenius, A., 1990
DESIGN OF SANDWICH PANELS AGAINST THERMAL LOADS, Proceedings of IABSE Conference, Brussels, Belgium, September, pp. 605–610.

Heagler, R. B., 1987
COMPOSITE FLOOR DECK IN TALL BUILDINGS, *Cold-Formed Steel,* Workshop Proceedings of 3rd International Conference on Tall Buildings, Council Report M370, January 6, 1986.

Heagler, R. B., 1989a
COMPOSITE FLOOR DECK, private correspondence.

Heagler, R. B., 1989b
LRFD DESIGN MANUAL FOR COMPOSITE BEAMS AND GIRDERS WITH STEEL DECKS, Steel Deck Institute, Canton, Ohio.
Heinzerling, J. E., 1971
STRUCTURAL DESIGN OF STEEL JOIST ROOFS TO RESIST PONDING LOADS, Technical Digest 3, Steel Joist Institute, Myrtle Beach, S.C., May.
Hetrakul, N., and Yu, W. W., 1978
STRUCTURAL BEHAVIOUR OF BEAM WEBS SUBJECTED TO WEB CRIPPLING AND A COMBINATION OF WEB CRIPPLING AND BENDING, Final Report, Civil Engineering Study 78-5, University of Missouri-Rolla, Rolla, Mo., June.
Hoff, N. J., and Mautner, S. E., 1956
THE BUCKLING OF SANDWICH-TYPE PANELS, *Journal of the Aeronautical Sciences,* vol. 12, pp. 285–297.
Höglund, T., 1986
LOAD BEARING STRENGTH OF SANDWICH PANEL WALLS WITH WINDOW OPENINGS, Proceedings of IABSE Colloquium, Stockholm, Sweden, May, pp. 349–356.
Hsiao, L. E., Yu, W. W., and Galambos, T. V., 1990
AISI LRFD METHOD FOR COLD-FORMED STEEL STRUCTURAL MEMBERS, *Journal of Structural Engineering, ASCE,* vol. 116, no. 2.
IISI, 1988
STEEL IN HOUSING, International Iron and Steel Institute, Brussels, Belgium.
ISO 898/I, 1987
MECHANICAL PROPERTIES OF FASTENERS, PART 1: BOLTS, SCREWS AND STUDS, International Standards Organization, Geneva.
Iyengar, S. H., and Zils, J. J., 1973
COMPOSITE FLOOR SYSTEMS FOR SEARS TOWER, *Engineering Journal,* American Institute of Steel Construction, Chicago, Ill., Third Quarter, pp. 74–81.
Johnson, A. L., 1966
THE STRUCTURAL PERFORMANCE OF AUSTENITIC STAINLESS STEEL MEMBERS, Report 327, Department of Structural Engineering, Cornell University, Ithaca, N.Y.
Johnson, A. L., and Kelson, G. A., 1969
STAINLESS STEEL IN STRUCTURAL APPLICATIONS, in *Stainless Steel for Architectural Use,* STP 454, American Society for Testing and Materials, Philadelphia, Pa.
Johnson, A. L., and Winter, G., 1966
BEHAVIOR OF STAINLESS STEEL COLUMNS AND BEAMS, *Journal of the Structural Division, ASCE,* vol. 92, October.
Karman, T. von, and Biot, M. A., 1940
MATHEMATICAL METHODS IN ENGINEERING, 1st edition, McGraw-Hill, New York, p. 304.
Kennedy, D. J. L., and Gad Aly, M., 1980
LIMIT STATES DESIGN OF STEEL STRUCTURES—PERFORMANCE FACTORS, *Canadian Journal of Civil Engineering,* vol. 7, March.
König, J., 1981
THE COMPOSITE BEAM ACTION OF COLD-FORMED SECTIONS AND BOARDS, Document D14: 1981, Swedish Council for Building Research, Stockholm.
Kozák, J., 1991
STEEL-CONCRETE STRUCTURES FOR MULTISTOREY BUILDINGS, in *Developments in Civil Engineering,* vol. 35, Elsevier, Amsterdam.
Kudder, R. J., Linehan, P. W., and Wiss, J. F., 1978
STATIC AND ULTIMATE LOAD BEHAVIOR OF COLD-FORMED STEEL-JOISTS RESIDENTIAL FLOOR SYSTEMS, Proceedings of 4th International Specialty Conference on Cold-Formed Steel Structures, University of Missouri-Rolla, Rolla, Mo., pp. 587–614.

Kuenzi, E. W., 1960
STRUCTURAL SANDWICH DESIGN CRITERIA, Publication 798, National Academy of Sciences–National Research Council, pp. 9–18.

LaBoube, R. A., and Yu, W. W., 1978a
STRUCTURAL BEHAVIOUR OF BEAM WEBS SUBJECTED TO BENDING STRESS, Final Report, Civil Engineering Study 78-1, University of Missouri-Rolla, Rolla, Mo., June.

LaBoube, R. A., and Yu, W. W., 1978b
STRUCTURAL BEHAVIOUR OF BEAM WEBS SUBJECTED PRIMARILY TO SHEAR STRESS, Final Report, Civil Engineering Study 78-1, University of Missouri-Rolla, Rolla, Mo., June.

Lawson, R. M., Mullet, D. L., and Ward, F. P. D., 1990
GOOD PRACTICE IN COMPOSITE FLOOR CONSTRUCTION, Publication 090, Steel Construction Institute, London.

Lin, S. H., 1989
LOAD AND RESISTANCE FACTOR DESIGN OF COLD-FORMED STAINLESS STEEL, Ph.D Thesis, University of Missouri-Rolla, Rolla, Mo.

Lin, S. H., Yu, W. W., and Galambos, T. V., 1988
ASCE STANDARD FOR STAINLESS STEEL STRUCTURES, Proceedings of 9th International Specialty Conference on Cold-Formed Steel Structures, University of Missouri-Rolla, Rolla, Mo., pp. 681–696.

Linehan, P. W., Kudder, R. J., and Wiss, J. F., 1978
DYNAMIC AND HUMAN RESPONSE BEHAVIOR OF COLD-FORMED STEEL-JOISTS RESIDENTIAL FLOOR SYSTEMS, Proceedings of 4th International Specialty Conference on Cold-Formed Steel Structures, University of Missouri-Rolla, Rolla, Mo., pp. 615–646.

Lula, R. A., 1984
THE STATUS AND DEVELOPMENT AND INDUSTRIAL APPLICATION OF THE 12% CHROMIUM STEELS IN THE UNITED STATES, Proceedings of the Inaugural International 3CR12 Conference, Middelburg Steel and Alloys, Johannesburg, Republic of South Africa.

Luttrell, L. D., 1980
STEEL DECK INSTITUTE DIAPHRAGM DESIGN MANUAL, Steel Deck Institute, Canton, Ohio.

Luttrell, L. D., 1981
STEEL DECK INSTITUTE DIAPHRAGM DESIGN MANUAL, 1st edition, St. Louis, Mo.

Luttrell, L. D., 1986
METHODS FOR PREDICTING STRENGTH IN COMPOSITE SLABS, Proceedings of 8th International Specialty Conference on Cold-Formed Steel Structures, University of Missouri-Rolla, Rolla, Mo., pp. 419–432.

Luttrell, L. D., 1991
ROOF AND FLOOR DIAPHRAGMS, private correspondence, May 1.

Luttrell, L. D., and Davison, J. H., 1973
COMPOSITE SLABS WITH STEEL DECK PANELS, Proceedings of 2d Specialty Conference on Cold-Formed Steel Structures, University of Missouri-Rolla, Rolla, Mo., pp. 573–601.

Luttrell, L. D., and Huang, H. T., 1980
THEORETICAL AND PHYSICAL EVALUATIONS OF STEEL SHEAR DIAPHRAGMS, Proceedings of 5th International Specialty Conference on Cold-Formed Steel Structures, University of Missouri-Rolla, Rolla, Mo.

Marchello, M. J., 1989
STEEL-FRAMED WALL APPLICATIONS, private correspondence.

McCraig, L. A., and Schuster, R. M., 1988
REPEATED POINT LOADING ON COMPOSITE SLABS, Proceedings of 9th International Specialty Conference on Cold-Formed Steel Structures, University of Missouri-Rolla, Rolla, Mo., pp. 361–386.

Melvill, M. L., Mahony, C. S., Hoffman, J. P., and Dewar, K., 1980
THE DEVELOPMENT OF A CHROMIUM-CONTAINING CORROSION RESISTING STEEL, presented at the 3d South Africa Corrosion Conference.
Merwe, P. van der, 1987
DEVELOPMENT OF DESIGN CRITERIA FOR FERRITIC STAINLESS STEEL STRUCTURAL MEMBERS AND CONNECTIONS, Ph.D Thesis, University of Missouri-Rolla, Rolla, Mo.
Merwe, P. van der, and Yu, W. W., 1985
DEVELOPMENT OF DESIGN CRITERIA FOR A FERRITIC STAINLESS STEEL, Structural Engineering Congress, ASCE Structural Division, Chicago, Ill.
Miller, C. J., 1973
DRIFT CONTROL WITH LIGHT GAGE STEEL INFILL PANELS, Proceedings of 2d Specialty Conference on Cold-Formed Steel Structures, University of Missouri-Rolla, Rolla, Mo., pp. 437–466.
Miller, C. J., 1974
LIGHT GAGE STEEL INFILL PANELS IN MULTISTORY STEEL FRAMES, *Engineering Journal,* American Institute of Steel Construction, Chicago, Ill., pp. 42–47.
NBCC, 1985
NATIONAL BUILDING CODE OF CANADA, 1985, National Research Council of Canada, Ottawa, Ont.
Newman, J. H., 1966
THE DRY FLOOR—A NEW APPROACH TO HIGH RISE APARTMENT BUILDINGS, *Engineering Journal,* American Institute of Steel Construction, Chicago, Ill., pp. 161–164.
Nowak, A. S., and Lind, N. C., 1979
PRACTICAL CODE CALIBRATION PROCEDURES, *Canadian Journal of Civil Engineering,* vol. 6, March.
Parimi, S. R., and Lind, N. C., 1976
LIMIT STATES BASIS FOR COLD-FORMED STEEL DESIGN, *Journal of the Structural Division, ASCE,* vol. 102, March.
Pekoz, T., and McGuire, W., 1979
WELDING OF SHEET STEEL, Research Report SG 79-2, AISI, January.
Phung, N., and Yu, W. W., 1978
STRUCTURAL BEHAVIOUR OF TRANSVERSELY REINFORCED BEAM WEBS, Final Report, Civil Engineering Study 78-5, University of Missouri-Rolla, Rolla, Mo., June.
Porter, M. L., 1986
HIGHLIGHTS OF NEW ASCE STANDARD ON COMPOSITE SLABS, Proceedings of 8th International Specialty Conference on Cold-Formed Steel Structures, University of Missouri-Rolla, Rolla, Mo., pp. 433–452.
Porter, M. L., 1988
TWO-WAY ANALYSIS OF STEEL DECK FLOOR SLABS, Proceedings of 9th International Specialty Conference on Cold-Formed Steel Structures, University of Missouri-Rolla, Rolla, Mo., pp. 331–342.
Porter, M. L., and Ekberg, C. E., 1976
DESIGN RECOMMENDATIONS FOR STEEL DECK FLOOR SLABS, *Journal of the Structural Division, ASCE,* vol. 102, pp. 2121–2136.
Porter, M. L., and Greimann, L. F., 1982
COMPOSITE STEEL DECK DIAPHRAGM SLABS—DESIGN MODES, Proceedings of 6th International Specialty Conference on Cold-Formed Steel Structures, University of Missouri-Rolla, Rolla, Mo., pp. 467–484.
Porter, M. L., and Greimann, L. F., 1984
SHEAR-BOND STRENGTH OF STUDDED STEEL DECK SLABS, Proceedings of 7th International Specialty Conference on Cold-Formed Steel Structures, University of Missouri-Rolla, Rolla, Mo., pp. 285–306.

prEN 10 147, 1989
CONTINUOUSLY HOT-DIP ZINC COATED UNALLOYED STRUCTURAL STEEL SHEET AND STRIP; TECHNICAL DELIVERY CONDITIONS, European Committee for Standardization, CEN, Brussels, Belgium.
RMI, 1980
INDUSTRIAL STEEL STORAGE RACK MANUAL, Rack Manufacturers Institute, Pittsburgh, Pa.
Robinson, H., 1969
COMPOSITE BEAM INCORPORATING CELLULAR STEEL DECKING, *Journal of the Structural Division, ASCE,* vol. 95, pp. 355–380.
Robinson, H., Fahmy, E. H., and Azmi, M. H., 1978
COMPOSITE OPEN-WEB JOISTS WITH FORMED METAL FLOOR, *Canadian Journal of Civil Engineering,* vol. 5, pp. 1–10.
Schmidt, L. C., and Roach, D. A., 1973
LATERAL BUCKLING OF OPEN-WEB JOISTS, *Institute of Engineering—Australian Civil Engineering Transactions,* vol. CE, no. 1, pp. 27–33.
Schurter, P. G., and Schuster, R. M., 1986
ALUMINUM-ZINC ALLOY COATED STEEL FOR COMPOSITE SLABS, Proceedings of 8th International Specialty Conference on Cold-Formed Steel Structures, University of Missouri-Rolla, Rolla, Mo., pp. 487–508.
SDI, 1987a
DESIGN MANUAL FOR COMPOSITE DECKS, FORM DECKS AND ROOF DECKS, Steel Deck Institute, Canton, Ohio.
SDI, 1987b
STEEL DECK INSTITUTE DIAPHRAGM DESIGN MANUAL, 2d edition, Steel Deck Institute, Canton, Ohio.
SDI, 1987c
DECK DAMAGE AND PENETRATIONS, Steel Deck Institute, Canton, Ohio.
SDI, 1989
DESIGN MANUAL, Publication 27, Steel Deck Institute, Canton, Ohio.
Seleim, S. S., and Schuster, R. M., 1982
SHEAR-BOND CAPACITY OF COMPOSITE SLABS, Proceedings of 6th International Specialty Conference on Cold-Formed Steel Structures, University of Missouri-Rolla, Rolla, Mo., pp. 511–532.
SIS, 1991
COMPOSITE SLABS, PRACTICE ASPECTS (in Dutch), Samenwerkingsverband Industrie Staalplaat-betonvloeren (SIS), TNO-Building and Construction Research Report BI-90-192, Rijswijk, The Netherlands.
SJI, 1988
STANDARD SPECIFICATIONS, LOAD TABLES, AND WEIGHT TABLES FOR STEEL JOISTS AND JOIST GIRDERS, Steel Joist Institute, Myrtle Beach, S.C.
Sokolnikoff, I. S., 1956
MATHEMATICAL THEORY OF ELASTICITY, 2d edition, McGraw-Hill, New York, pp. 249–262.
Somers, R. E., and Galambos, T. V., 1983
WELDING OF OPEN WEB STEEL JOISTS, Technical Digest 8, Steel Joist Institute, Myrtle Beach, S.C., August.
Sprague, E. T. E., 1978
50-YEARS DIGEST—A COMPILATION OF SJI SPECIFICATIONS AND LOAD TABLES FROM 1928 TO 1978, Technical Digest 7, Steel Joist Institute, Myrtle Beach, S.C., November.
Stark, J. W. B., 1978
DESIGN OF COMPOSITE FLOORS WITH PROFILED STEEL SHEET, Proceedings of 4th International Specialty Conference on Cold-Formed Steel Structures, University of Missouri-Rolla, Rolla, Mo., pp. 893–922.

Stark, J. W. B., and Brekelmans, J. W. P. M., 1990
PLASTIC DESIGN OF CONTINUOUS SLABS, *Journal of Constructional Steel Research,* vol. 15, pp. 23–47.
Temple, M. C., and Abdel-Sayed, G., 1978
FATIGUE EXPERIMENTS ON COMPOSITE SLAB FLOORS, Proceedings of 4th International Specialty Conference on Cold-Formed Steel Structures, University of Missouri-Rolla, Rolla, Mo., pp. 871–892.
Tham, L. G., Chong, K. P., and Cheung, Y. K., 1982
FLEXURAL BENDING AND AXIAL COMPRESSION OF ARCHITECTURAL SANDWICH PANELS BY COMBINED FINITE-PRISM-STRIP METHOD, *Journal of Reinforced Plastics and Composite Materials,* vol. 1, pp. 16–28.
Thomas, C. R., 1981
USE OF A 12% CHROMIUM FERRITIC STAINLESS STEEL IN COAL CONVERSION SYSTEMS, in *Corrosion Resistant Materials for Coal Conversion Systems,* D. B. Meadow and M. I. Manning (Eds.), Essex, England: Applied Science Publishers, pp. 259–271.
Tide, R. H. R., and Galambos, T. V., 1970
COMPOSITE OPEN-WEB STEEL JOISTS, *Engineering Journal,* American Institute of Steel Construction, Chicago, Ill., pp. 27–36.
Timoshenko, S., and Goodier, J. N., 1951
THEORY OF ELASTICITY, 2d edition, McGraw-Hill, New York, chapters 2 and 3.
Tomà, A. W., and Stark, J. W. B., 1982
CONNECTIONS IN THIN-WALLED STRUCTURES, in *Developments in Thin-Walled Structures,* vol. 1, J. Rhodes and A. C. Walker (Eds.), Essex, Applied Science Publishers, England, chapter 5, pp. 159–203.
Tomasetti, R. L., Gutman, A., Lew, I. P., and Joseph, L. M., 1986
DEVELOPMENT OF THIN WALL CLADDING TO REDUCE DRIFT IN HI-RISE BUILDINGS, IABSE Colloquium, Stockholm, Sweden, May, ETH-Hönggerberg, Zürich, Switzerland.
Trestain, T. W. J., 1982
A REVIEW OF COLD FORMED STEEL COLUMN DESIGN, Report 81109-1, CSCC Project CSS817, December.
TRI, 1973
SEISMIC DESIGN FOR BUILDINGS, Tri-Services Manual, U.S. Army Technical Manual 5-809-10, Department of the Army, Washington, D.C., April.
TRI, 1982
SEISMIC DESIGN FOR BUILDINGS, Tri-Services Manual, Army, Navy and Air Force, U.S. Government Printing Office.
Wang, S. T., 1969
COLD-ROLLED AUSTENITIC STAINLESS STEEL: MATERIAL PROPERTIES AND STRUCTURAL PERFORMANCE, Report 334, Department of Structural Engineering, Cornell University, Ithaca, N.Y.
Wang, P. C., and Kaley, D. J., 1967
COMPOSITE ACTION OF CONCRETE SLAB AND OPEN WEB JOIST (WITHOUT THE USE OF SHEAR CONNECTORS), *Engineering Journal,* American Institute of Steel Construction, Chicago, Ill., pp. 10–16.
Wing, B. A., 1981
WEB CRIPPLING AND THE INTERACTION OF BENDING AND WEB CRIPPLING OF UNREINFORCED MULTI-WEB COLD-FORMED STEEL SECTIONS, M.A.Sc. Thesis, Department of Civil Engineering, University of Waterloo, Waterloo, Ont., Canada.
Yu, W. W., 1991
COLD-FORMED STEEL DESIGN, New York: John Wiley.
Zienkiewicz, O. C., and Cheung, Y. K., 1967
THE FINITE ELEMENT METHOD IN STRUCTURAL AND CONTINUUM MECHANICS, McGraw-Hill, New York.

Contributors

The following list acknowledges those who have contributed materials for this Monograph. The names, affiliations, and countries of each contributor are given.

Rolf Baehre, University Fridericana, Karlsruhe, Germany
Jan Brekelmans, TNO Building and Construction Research, Rijswijk, The Netherlands
Byron Daniels, TNO Building and Construction Research, Rijswijk, The Netherlands
Ken P. Chong, National Science Foundation, Washington, D.C., USA
Steven R. Fox, Canadian Sheet Steel Building Institute, Willowdale, Ont., Canada
Richard B. Heagler, Nicholas J. Bouras, Inc., Summit, N.J., USA
Larry D. Luttrell, West Virginia University, Morgantown, W.V., USA
Pieter van der Merwe, Rand Afrikaans University, Johannesburg, Republic of South Africa
Reini M. Schuster, University of Waterloo, Waterloo, Ont., Canada
Jan Stark, TNO Building and Construction Research, Rijswijk, and Technical University Eindhoven, Eindhoven, The Netherlands
Derek L. Tarlton, Canadian Sheet Steel Building Institute, Willowdale, Ont., Canada
Ton Tomà, TNO Building and Construction Research, Rijswijk, The Netherlands
Don S. Wolford, Private Consultant, Middletown, Ohio, USA
Douglas Yates, General Motors of Canada, St. Catherines, Canada
Wei-Wen Yu, University of Missouri-Rolla, Rolla, Mo., USA

Name Index

Subject Index